Becoming Muslim

Culture, Mind, and Society

The Book Series of the Society for Psychological Anthropology

With its book series Culture, Mind, and Society and journal *Ethos*, the Society for Psychological Anthropology publishes innovative research in culture and psychology now emerging from the discipline of anthropology and related fields. As anthropologists seek to bridge gaps between ideation and emotion or agency and structure—and as psychologists, psychiatrists, and medical anthropologists search for ways to engage with cultural meaning and difference—this interdisciplinary terrain is more active than ever.

This book series from the Society for Psychological Anthropology establishes a forum for the publication of books of the highest quality that illuminate the workings of the human mind, in all of its psychological and biological complexity, within the social, cultural, and political contexts that shape thought, emotion, and experience.

Contents

Acknowledgments

This project would not have been realized without the insights and assistance of many people. I am indebted to those who have guided me through the journey of this book.

I am deeply grateful to the women who have shared their experiences and memories with me. Their words bring the vital substance to the study. Gunnar Alsmark, my mentor, inspired me to explore the field of European Ethnology and I owe him much. I am grateful to my friend and colleague Kristina Gustafsson who has read different versions and offered her thoughtful remarks. I have very much enjoyed her and Ingrid Fioretos' friendship and cheerful and constructive discussions over the years. Daniel Linger introduced me to psychological anthropology; I have benefited greatly from his invaluable comments and guidance. His perspective has clearly influenced the ideas and arguments in this book. For productive remarks and congenial conversations I also owe my thanks to Fredrik Nilsson, Lynn Åkesson, Orvar Löfgren, and Pernilla Ouis. I am indebted to my editor Douglas Hollan for his encouragement and helpful suggestions. Two anonymous reviewers offered much appreciated and invaluable comments. For financial support, Lund University assisted the project through various funds. The American-Scandinavian Foundation and the Swedish Institute generously funded my research in the United States.

Matt offered his remarkable optimism and patience during periods of doubt and stress. His dedicated support has always been there when I needed it the most. During the last stage of getting the book done, Alexander was born. He has brought us much joy and a vital perspective on life.

Becoming Muslim

WESTERN WOMEN'S CONVERSIONS TO ISLAM

Anna Mansson McGinty

First published in hardcover in 2006 by PALGRAVE MACMILLAN® in the United States—a division of St. Martin's Press LLC, 175 Fifth Avenue, New York, NY 10010.

Where this book is distributed in the UK, Europe and the rest of the world, this is by Palgrave Macmillan, a division of Macmillan Publishers Limited, registered in England, company number 785998, of Houndmills, Basingstoke, Hampshire RG21 6XS.

Palgrave Macmillan is the global academic imprint of the above companies and has companies and representatives throughout the world.

Palgrave® and Macmillan® are registered trademarks in the United States, the United Kingdom, Europe and other countries.

ISBN: 978–0–230–61668–4

Library of Congress Cataloging-in-Publication Data is available from the Library of Congress.

A catalogue record for this book is available from the British Library.

Design by Newgen Imaging Systems (P) Ltd., Chennai, India.

First PALGRAVE MACMILLAN paperback edition: September 2009

10 9 8 7 6 5 4 3 2 1

Printed in the United States of America.

Transferred to Digital Printing 2009

For my parents,
Åsa and Wiking

Part I
The Individual Conversion

Chapter One

Conversion and Identity Formation

I still have my friends from high school. I talk to them once a year. And they say, "Are you still Muslim? Are you raising your kids Muslim?" They think it is some kind of New Age fad that I might drop at any time, you know [laughs]. The problem in America, I don't know in Sweden, is religious people. Nonreligious people just don't understand how a smart, intelligent person can become religious. Even my mother, who had been searching all her life but had finally decided that there was nothing there, she just couldn't understand . . . They think of it as a weakness or crutch, and especially for an intelligent woman "why Islam?". . . I hear the same from Egyptian women—why do we always have to talk about the scarf when we talk about Islam? Why, if I put a piece of material on my head, do I become stupid? Why does this even matter for discussion! What do you care what somebody wears if what's inside doesn't change? (Mariam)

What I find difficult in my life is that since I'm wearing the veil I am questioned. I find it very hard that my intellect is questioned. You can't be intellectual or intelligent and Muslim at the same time. That combination does not exist for most Swedes. In their eyes I have to be somewhat stupid if I, as a Swedish woman, decide to convert to Islam (Marianne).

Mariam from the United States and Marianne from Sweden express disappointment and frustration about being misjudged and misunderstood as Western female converts[1] to Islam. They feel a distressing gap between their own self-image and the negative stereotypes many Americans and Swedes have of religious people in general and Muslim women in particular. They want to be valued for who they are, to be taken seriously. Each understands her conversion as an emotional and intellectual development in her own life, not as a compensation for weakness or a rejection of rationality.

In this book I depart from interviews that I have conducted with some women in Sweden and in the United States who have converted to Islam. I was interested in their identity and life as Muslim and what they found appealing within the Islamic faith. The women told me about particular experiences that were charged with emotional and existential meaning, such as having a dream, a separation, or going on trips to foreign places, experiences that had triggered a need for spiritual connection and fascination with another way of living. Many of them told me about a long process of reflecting, reading, discussing, and slowly trying out what was for them a new and different worldview. The conversions reflect in this sense a slow, gradual transformation. The accounts of the women, as well as others',[2] reflect a longing for a change of self, and over time the religious belief has been given salient meaning within a specific subjective world.

When I have talked about my research, in academic as well as nonacademic contexts, many people have ventilated skepticism and wonder about how Western women at all could deliberately convert to Islam, a religion they often associate with oppression of women. Overall, Islam has become a charged and widely discussed subject that generates a lot of emotions; everyone seems to have an opinion about it. The presence of Islam in the West is palpable, being the fastest growing religion in both Europe and the United States. Muslims are far from being the far-away "others." The often-invoked dichotomy and conflict between "West" and "Islam" is unfounded and questionable in some people's everyday life though at the same time it is being used politically to reinforce differences between an "us" and a "them." While Islam is frequently presented by media as a threat to "Western" ideas, some people in the West explore and commit themselves to the Muslim faith, finding in it support for ideas about solidarity, family, social justice, and peace. In light of this, and particularly after the terrorist attacks on September 11, 2001, experiences and narratives such as those of the female converts, reflecting nuances and identification with seemingly incompatible worldviews and cultural ideas, are of the utmost importance.

Anthropologists and social scientists in general, have for quite some time questioned the notion of cultures as bounded, timeless, and coherent wholes. Still, we frequently encounter accounts and reports on how people, for example, immigrant youth, are trapped *between* cultures. Individuals experience conflicting loyalties and internal conflicts between different cultural worldviews and beliefs,[3] but cultures or civilizations per se do not really clash. The converts experience conflicts with family members and awkward encounters with the surrounding world, but in terms of their sense of self as Swedish Muslims or American Muslims another equally important understanding of different cultural and religious ideas is highlighted in their narratives, an understanding that deserves our attention.

Through an arresting self-reflective process of meaning-making, seriously confronting any notion of culture as something outside people's minds, the female converts reconcile old and new, making Islam their own.

Looking back on my own life experiences, there might also be other more subtle reasons for my interest in Islam in the West. The first classes I attended as an undergraduate were in history of religion and science of religion. Even if religious matters are not essential parts of my own personal life or identity, I am fascinated by existential questions concerning the spiritual aspects of people's sense of self and their role in and impact on people's everyday life.

My first personal encounter with and experience of Islam and Muslims took place in the land of the two rivers, the Euphrates and the Tigris. At the age of fourteen I landed, together with my parents and sister, at the airport in Baghdad. This was the beginning of a one-year visit to Iraq. I remember very well the warm, dry air and the sound of the wind in the palm leaves when I stepped out of the plane. It was a dark, clear night with lots of stars and a crescent moon. I also remember waking up early the next morning, somewhat confused as to where I was, hearing the calling from a nearby minaret. It was undeniably another world that I encountered in my early adolescence. Soon this world, with the *souqs*,[4] the neighbor shops, and the bakery around the corner from where we lived, the soldiers on the rooftops, the walk to the pool, became parts of my everyday life. At Baghdad International School I made friends, in particular two good female friends, Ayşe from Turkey and Sumera from Pakistan. They were from two different parts of what is frequently referred to as the "Muslim world" and they had quite different approaches to Islam. Besides teenage conversations about pop groups and boys, we sometimes talked about religion, especially Sumera and I. Ayşe was from a secular family where religion was not an integral part of life, while Sumera, veiled and modestly dressed, was a practicing Muslim, raised in a devoted religious family. I remember, for example, that she was not allowed by her father to celebrate *Lucia*[5] with me and some other Swedish families. My own traditions were then placed and understood in a religious context seen from someone else's perspective. I mention this since that year, I believe, had an impact on me in different ways and when reflecting on my research interests this experience stands out in my own reconstructed life story. I often get the question why I chose to study conversion to Islam, and I believe there are, just as in the case of conversion, many different contributing factors, some easier to communicate than others.

Important for Research

Change and Continuity of Self

Conversion triggers profound questions to the self; it heightens the awareness and prompts reflections of who one is, who one was, and where one is heading. It captures both popular and scholarly interest since it raises critical problems regarding human beings' reflections on the meaning of life and their relation to a metaphysical sphere. For me as an anthropologist, the phenomenon of religious conversion poses crucial questions regarding identity and how people make sense of changes in life. How do people change? And what is it that really has changed? This book is about the identity formation of some female Muslim converts; it concerns the personal meaning of becoming Muslim. The overarching purpose is to analyze self-making, the transformation as well as continuity of the self, in light of appropriating a religious belief and integrating it into a personal life and of communicating the change of identity to the surrounding world.

It should be made clear that as an anthropological study, this work does not problematize conversion from a religious perspective but as a cognitive and cultural phenomenon, though with the intention to approach phenomenologically the experience of the convert. It is not primarily a book about different Islamic beliefs or Islam as a religious system but about the particular individuals who embrace the religion, who talk about it and understand it in various ways, and the process of meaning-making this entails.

In a recent anthropological anthology on conversion (Buckser and Glazier 2003), several excellent ethnographic case studies are offered but none of them explores conversion to Islam. As psychologist of religion Lewis R. Rambo (2003) points out in the afterword, more work within this area needs to be done particularly with the growing number of Muslims around the world and the political situation after September 11, 2001 in mind. While extensive research has been conducted on Muslims in the West, fewer studies have been done on conversion to Islam within the social sciences in general. Two studies written from a historical standpoint are *Conversion to Islam in the Medieval Period* (Billiet 1979) and *Conversion to Islam* (Levtzion 1979), the latter focusing on other geographical contexts than the West. In contrast, Larry Poston, a religious scholar, aims to address this lack in his *Islamic Da'wah in the West: Muslim Missionary Activity and the Dynamics of Conversion to Islam* (1992). His work offers a thorough and interesting analysis of *da'wah*, Muslim missionary activity, in the West; but his discussion of the testimonies of American and European converts, based on questionnaires and testimonies published in various Islamic journals, is much less exhaustive.[6] Another more recent study about female American converts, also based on questionnaires, is Carol Anway's [*Daughters of Another Path: Experiences of American Women Choosing Islam*] (1996, see also 1998). Using surveys, both studies focus on overall similarities between the accounts and general trends and themes, some of them recognizable in my own research material. On the contrary, Ali Köse's (1996) study is based on interviews with native British converts to Islam, but like the studies mentioned above, his focus is on the major factors involved in their conversions and why they chose Islam.[7] Neither of the studies offers an indepth analysis of subjective experiences of conversion and the intricate processes of identity formation prompted by conversion.

The experiences of the converts demonstrate a personally and socially recognized transformation of the self. The internalization of a religious belief system brings about major changes in the individual's subjective reality, but in the same process, she simultaneously maintains a strong sense of continuity. People strive to create coherence across all major changes and interruptions in their lives so as to render life meaningful. This problem pinpoints some crucial questions. What triggers conversion and what characterizes the appropriation of a religious belief? How do people make sense of diverse cultural and religious ideas as well as a break in worldview? What distinguishes the transformation and how can we explain the simultaneous experience of self-continuity? How exactly does the self transform and organize coherence of several distinct cultural messages?

Throughout the book I highlight the microprocesses of identity-making more than the macro-processes and larger social contexts. Above all, the life stories of the women reflect the diverse ways they had become Muslims and the idiosyncratic resonance of the religious system. By analyzing conversion as a psychocultural process, I will demonstrate how a convert integrates and reconciles various ideas and representations within her particular "personal world."[8] Her narratives illustrate the interplay between public representations and personal appropriations, not only of Islam but also of gender issues and other social and political ideas. This psychological appropriation of discourses deserves, I believe, more attention from anthropologists, and it compels us to raise several questions: How does each convert incorporate Islam into her personal identity? What does Islam mean to each? Which emotions and memories are attached to the conversion experience? How do ideologies and cultural messages and symbols become meaningful? In what ways can we explain people's sense of coherence despite the appropriation of multiple and seemingly irreconcilable discourses?

Building on one of the main ideas within psychological anthropology that we have to look at the internalization of culture, this study demonstrates how, by giving attention to the personal meaning assigned to discourses and the emotions invested in them, one can better understand why and how some discourses and systems of ideas gain strong ideological force and why certain structures are reproduced (cf. Strauss and Quinn 1997). The longing for change in the self, expressed by the converts, suggests not only personal desires but also a critical commentary on discursive ideas around them. I argue then, together with other psychological anthropologists, that by acknowledging a psychological dimension in the study of culture we can gain a greater understanding of cultural variation, change, and reproduction. This requires us to attend to an often neglected realm—the reflective sense of self.

Meaning-Makers

Becoming Muslim is experienced and told in a specific sociocultural context, but always infused by idiosyncratic experiences and memories. The women are not simply shaped by the interactions nor do they perform the particular "positions" ascribed to them; they engage representations, categories, and people around them, including myself (cf. Linger 2001). Their interactions with me (the interviews) not only "force" them to verbalize experience and to make sense of why they converted, but they also made me think about the meaning of their presentation and the meaning that the converts, me included, assign to categories such as "Swedish" and "Muslim." An essential aspect is stressed throughout the study: the aspect of *doing*, so apparent in the process of meaning-making, that takes place in the interplay between the realm of cognition and feeling and the realm of discourses. This feature of "doing" by the self permeates the discussions on narratives as transitional (part I), on personal models (part II), and on "looping effects" of meaning (part III).

In his recent work, Linger (2005) urges anthropologists to carefully think through the model of the person that underlies their work and of the consequences it has on assumptions of meaning and identity. The theoretical approach of personal models explored in this book rests on a certain model of self that stresses the women's self consciousness, that is, reflective awareness of themselves, agency, and their active engagement with others and available categories (cf. Cohen 1994; Linger 2001, 2005; Sökefeld 1999). Advocating a similar view does not imply that one understands the self as autonomous. The self is unmistakably social and cultural, but not a passive subject of society and culture. Individuals are meaning-makers, or in Anthony Cohen's words, "world-makers," who make the external world theirs through "their acts of perception and interpretation" (Cohen 1994:115). This book offers examples of how the women assign *personal meaning* to a religious belief and make it their own (Obeyesekere 1981, Stromberg 1993). The convert is trying to make sense of herself in the world by employing and integrating diverse and sometimes seemingly irreconcilable representations. Throughout the conversion process the women reconcile old with new, reflecting the process of meaning-making. It is hard to ignore this agency[9] and reflexivity, which are required to be able to embrace and negotiate diverse messages and self-representation and at the same time sense some coherence. Consequently, neither sense of self nor meaning is fixed; both are objects of reflection and transformation.

Personal Models

The proposed model of the person, as well as the assumption that culture is not equally reflected in public and mental representations, entails analytical consideration of the internal structures through which individuals internalize and understand discourses.

A main point that I want to make in this book is that conversion is triggered and organized by *personal models.* [Conversion is not primarily incited and guided by dominant discourses, language, or social structures, but rather prompted by particular personal beliefs, quests, and desires.] Islam appeals to the women since its religious ideas address certain needs and already existing thoughts (cf. Sachs Norris 2003). The step toward "the unknown" is compelled by what I will describe later as *cognitive recognition.*

So what do I mean by personal models? Drawing on the work of psychological/cognitive anthropologists,[10] these models are internal structures of thought and feeling, through which people acquire and organize their understanding about themselves and the world. [I argue that in the process of becoming Muslim, the women appropriate and understand Islam and reformulate the self through particular cognitive models.] The mental structures, themselves the products of previous interactions with the public world, are as Claudia Strauss describes them, "learned, internalized patterns of thought-feeling that mediate both the interpretation of ongoing experience and the reconstruction of memories" (Strauss 1992a:3). These personalized patterns of thought-feeling[11] not only have emotional salience and motivational force but are also strongly linked to the women's self-understanding (cf. Quinn 1992). Personal models embody essential aspects of a person's inner world, guiding her understanding of the world and her place in it. Also, since they trigger the conversion itself, they have strong motivational force (D'Andrade 1985, D'Andrade and Quinn 1992).

A great deal of research within psychological anthropology, of which cognitive anthropology[12] is a part, has been done on cognitive models referred to as "cultural models," implying shared understandings within a group. These may be, to mention some, the models behind Americans' talk about gender types and romance (Holland and Skinner 1987, Holland 1992), of marriage (Quinn 1992), of success (Strauss 1992b), and of the self (D'Andrade 1987).[13] While these and other studies look at cultural models, the personal models that I elaborate here are of a more idiosyncratic nature. Both personal and cultural models are cognitive/emotional phenomena, but whereas personal models are subjective phenomena (or an individual organization of cognitive models that Schwartz (1978) refers to as "idioverses") cultural models are shared personal models. My point is that the converts internalize[14] and understand Islam by drawing from quite different personal set of ideas and knowledge infused with subjectivity, emotions, and desires, and linked to specific memories. These understandings and knowledge may be loosely inferred to cultural ideas around them, but most significantly they are intimately interwoven with biographical idiosyncrasies. I will show how these personal models have become and remain salient because they are attached to particular important ideas and emotional experiences and memories from childhood and adolescence (or later) and are reinforced, but also modified, by Muslim ideas and new experiences acquired through the process of conversion.

Claudia Strauss offers a similar analysis of idiosyncratic systems of meaning "shaped by the particular combination of experiences and ideas to which each individual is exposed, which are never exactly the same for any two people" (1992b:220). In her article about five blue-collar workers' talk about "getting ahead," she acknowledges, besides a shared model of success, what she calls, "personal semantic networks," which are linked to key symbols, life experiences, and conscious self-understandings (ibid.:211). As with the personal models explored here, these personal networks have stronger directive force than some cultural models, in Strauss' case the widely shared ideas of success.

There are other anthropological studies that have fruitfully ana-lyzed conversion as a continuous process rather than a radical, sudden change (Buckser and Glazier 2003) as well as demonstrated the psychological dimension of conversion.[15] In her fascinating account on magic and conversion to Wicca, Luhrmann (1989) elaborates around a thesis she calls "interpretative drift," a sociocognitive process, to explain how people are drawn to witchcraft and become practitioners and magicians themselves.[16] Compared to her model, which stresses the unmotivated and the not deliberate aspects of this process, the cognitive process studied here derives from the conscious thought and reflection of the converts. This study is about the converts' conscious identity formation and reflections upon continuity and change and rests on the assumption that personal models and unspoken knowledge can become objects of reflection and conscious thought (Strauss and Quinn 1997:46), something which is triggered by the conversion itself. Similarly, Stromberg (1990, 1993), in his work on conversion to Evangelical Christianity, is interested in the convert's sense of transformation and the personal meaning given to a religious belief. But rather than looking at internal structures, he offers a compelling analysis of how the believer brings up emotional conflicts and tries to resolve these through the very use of, and the personal meaning given to, the language of Evangelical Christianity. By reframing personal experiences and emotions through a canonical language the believer also gains a sense of self-transformation.[17]

I aspire to show how adopting the theory of personal models on the study of religious conversion offers a useful approach in several regards. It proposes a way to understand the simultaneous processes of appropriating Islam and organizing a Muslim identity, how new ideas are integrated with an already existing cognitive framework, the concurrent sense of change and continuity, and why Islam is appealing to the women in the first place. It also gives scope for multiple personal interpretations and understandings of the same public representation and how cognitive/emotional salience is given to a belief or a set of ideas. In that regard it departs from many previous studies within the social sciences on conversion, which are done with quite an opposite ambition in mind, namely that of drawing a profile of the "typical" religious convert and searching for a universal model of religious conversions in general.[18] Moreover, it should be pointed out that the mental process demonstrated here is not necessarily specific to religious conversion but could most likely be found in other kinds of self-transformations as well. The advocated theoretical model could, I believe, be fruitfully applied not only to conversions such as political and ethnic conversions but also to profound identity changes such as gender change.[19] The cognitive/emotional change in these cases might very well be triggered and illuminated by significant personal models, but I leave that for other studies to explore. This also generates a more general question: Don't we all change throughout life? The way people make sense of change and reconcile different messages through conversion, what I refer to as cognitive reconciliation, seem to point to a central feature of meaning-making in general.

It should be stressed that by focusing on a few models I do not intend to reduce the women's self-image to single themes. Rather, I want to highlight the particular ideas, emotions, and experiences that the women engage with when talking about the conversion and their Muslim identity at the particular time of the interviews, reflecting the ongoing work of identity-making. Consequently, I explore only some aspects of the women's identities. The many interview excerpts included in this volume reflect the interview encounter as one of many different opportunities to organize a sensed self-coherence, despite life-changing experiences.

Neither, should the personal models be understood as static and rigid, but as flexible ones that the individual can challenge, reinforce, or change when new ideas are internalized or through certain practices. Some cognitive anthropologists have shown the connection between cognitive models and motivation. Since models mediate certain views of the world, they also often offer desirable and advantageous goals to the person (D'Andrade and Strauss 1992). Once Islam is acquired as a belief with a high level of cognitive salience it will most likely guide or even instigate action (Spiro 1987 [1982]:164). But as Luhrmann (1989) so nicely demonstrates, belief can be motivated and be guided by action itself. I do not understand cognition as clearly separate from performances, but rather as intimately linked to human practices. As I will show, in one case the convert's beliefs were guided and learned through certain practices.

The idea of religious conversion as a cognitive/emotional process motivated by personal models is further explored through, what I refer to as, *cognitive recognition* and *reconciliation*, processes reflecting the self's capacity of meaning-making. The step toward "the unknown" is kindled by cognitive recognition, that is certain Islamic

ideas and practices are identified and given thought because they
address and appeal to a particular personal model, existing ideas and
desires. This recognition sparks fascination and personal engagement;
the ideas and practices make sense, offer appealing explanations,
and meet certain personal quests and emotions. In the process of
internalizing Islam, the religious ideas interact with, are examined
by, and merge with already internalized ideas about the world. When
Islam is gradually taken on, the personal model, through which the
belief is appropriated and understood, is both strengthened and
modified.

This study shows how *cognitive reconciliation* is a significant
feature of the whole conversion process. Notably, I believe, the
phenomenon of conversion renders itself as an excellent means
through which we can analyze the reflective and continuous integration
and reconciliation of diverse messages and representations into a
coherent "I," a quality characteristic for meaning-making in general.
Becoming Muslim does not really reach a final point. The converts,
and all of us for that matter, relate and negotiate our identities
continuously through life in the context of our new experiences. In
the case of the converts this means incorporating Islam into a personal identity and reconciling the break into a sensed continuity of
self and a coherent life story. Thus, cognitive reconciliation points
to the simultaneous processes of acquiring a religious belief and
self-transformation.

Beyond the biographical particularities of the women and their
diverse trajectory to Islam, the material points to an arresting notion
of the relation of conversion to self-making. The conversion itself,
and even the reflections on a potential conversion, seems to engender
a heightened sense of promise of exploring who one is and what
one's place in the world is. By approaching something new and
unknown the converts have a compelling and even liberating experience of how the conversion opens up for possibilities, for existential
elaboration of new identifications. One woman compared, on a symbolic level, her conversion to Islam with falling in love. It might be
hard to realize a desire for a change in life within and through
already well-known belief systems and domains. Through the exploration of something new or, as one convert put it, "tasting the flavor"
of Islam, the women experience new ways of rethinking themselves
and their place in the world. Leaving something old for something
yet unknown triggers an experience of anticipation that seemingly
anything is possible.

The Female Converts and I—Reflections on the "Field Encounter"

By placing the women at the center of inquiry, I seek to understand human experience as a central and dynamic part of social and cultural life. This kind of "person-centered ethnography," engaged in the subjective worlds of the women, aims to give "experience-near"[20] descriptions and analysis of human behavior and personal experience (Hollan 1997; 2000, Levy 1994, Linger 2001).

Below, I want to share some reflections regarding the material, method, and the fieldwork, since they give the reader important glimpses into how I went about this project as well as into the condition in the "field," which inevitably influenced the study. The material consists of a total of eighteen long, open-ended interviews with six Swedish women and three American women.[21] All interviews were collected between 1998 and 2001.[22] The study relies on these verbal accounts. In my discussion and analysis of the interviews, I draw from ideas, concepts, and emotions expressed by the women. That is, from the interviews I identify and interpret the knowledge, beliefs, and feelings that the women draw on in their understanding of Islam and in the work of self-making. My own research model then implies deriving and highlighting what seem to be significant understandings that are stressed and repeated during the interviews, offering a presentation of parts of the personal worlds of the women.

The problems of representation when using interview material are well acknowledged. There is the problematic relationship between narrative and experience, the relationship of thought and feeling to language, and the impact of the interview encounter. These factors are not to be ignored and will be dealt with here and further in chapter 3.

I do not believe it is possible to give an account of all the factors involved in the decision to convert or of all the aspects of the women's self-image for that matter. My analysis of personal models offers a thesis of what triggers the conversion and what the women find appealing in Islam. There is also a problematic relation between the event of conversion and the story about it. However, I presume that the self-narratives reflect prominent parts of subjective experience and a serious engagement and undertaking during the interview encounter to tell what really happened. We should, as Stromberg (1990:43) asserts, "take believers at their word that they have a sense of being transformed."

The names used here are of course pseudonyms. However, to assure a certain degree of anonymity, a firm request of the women, the use of pseudonyms was not enough in all cases. Particularly in regard to the women interviewed in Sweden, facts about background and particular recognizable distinctive features have in many cases been left out or slightly changed to reduce, as much as possible, the risk of identification. At the beginning of my "fieldwork" in Sweden I faced some problems. Due to mistrust and negative reactions to an article written by another social scientist in which some converts felt misrepresented, strong voices within the "convert community" persuaded female converts not to participate in any research, mine included. Later, one woman told me that she and the other women that did participate in my study were seen as betrayers. The fear and distress of being misunderstood and misrepresented was mentioned by a couple of the few women that did decline an interview. Other reasons were the tedium of all the attention and also a suspicion toward not only journalists but also researchers. What did I want? What was my purpose? What was so intriguing and fascinating about them and their convert friends?

I made acquaintance with the women through Muslim organizations but mostly by recommendation of the converts themselves. Particularly with regard to the interviews done in Sweden, this resulted in reaching a certain circle of converts. Rather than following a stricter religious approach to life, some of them preferred a more, as they referred to it, "liberal" application of the religion. This was explicitly manifested in some interviews; they called it a "Swedish Islam." Not all, but some of these women know each other and most of them have chosen rather public roles and are engaged in diverse activities and organizations. These are persons who *wanted* to tell their story and give a personal face to Islam. Some of the women will inevitably recognize each other's story even when particular biographical details have been removed. This particular fact that many of these converts know each other, and about each other, was for one woman the very reason for not participating in the study. She argued that regardless of "anonymity" and changes in facts, a certain network of converts would still recognize her and her life story.

Anne Sophie Roald has brought up the problematic aspects of studies and interviews done by non-Muslims on Muslims and the risks of misunderstanding and misconception that this can entail. She has rightly shown the disadvantages of an interview encounter when the interviewer and the informant do not share worldviews or "cultural language" (2001:70ff.). Without doubt, if I had been a Muslim or a man, or both, I would not have heard the same story and moreover, I would not have written the same book. Some of the women, particularly the ones interviewed in Sweden, may have undercommunicated the spiritual aspect of their conversion to me while emphasizing others. As Roald asserts, Muslims tend to speak of themselves in "worldly" terms in their interactions with non-Muslims. This was not the case of the American women though, who very much stressed the spiritual aspect of their conversion. If I had been a Muslim myself some of the women would probably not have felt as if they had to "rationalize" their decision or withhold from me strong religious feelings. On the other hand, I suspect that there were things that the women told me that they might not have told another Muslim due to judgment, expectations, and social control, concerning what they as Muslims are supposed to say and do. I do not share a common religious belief and commitment with the women I have interviewed, but it is my strong conviction that there are other crucial levels on which one can meet and from which understanding can emerge.

However, the relationship between my project and theirs is not unproblematic and the result of my analysis may not correspond to the women's own intentions. The women have had their own aims and purposes in participating in the study. With me being a young, female researcher, then in my late twenties, reaching out to female converts, they may have seen the interviews as a great opportunity to come out with their story. Besides being a sympathetic listener, I believe that I personally represented not only a research community but also, in some ways, a Swedish, Western, non-Muslim majority to which they very much wanted to reach out with their personal stories in a quest to challenge stereotypes about Islam and Muslims in general. Like members of other groups that are objects of negative representation and stereotypes, the women were very particular about giving me a positive and understandable picture in contrast to the preconceived notions of Islam and Muslims. In that sense I might, in some cases, have been given a "representative narrative," in which certain anxieties, doubts, and conflicts were undercommunicated.

The interview offers one of many different occasions to elaborate a feeling of coherence and to make sense of fundamental changes in life, for oneself as well as for others. The opportunity to be able to give reasonable and convincing explanations as to why one became a Muslim is most likely perceived as particularly important in relation to the surrounding world that often looks upon the women with suspicion and skepticism. Besides their wish to challenge distorted notions about Islam and Muslims, there could also be religious reasons for agreeing to be interviewed, doing it for God as a good Muslim. The converts may have directed their narrative to God, outside the present moment, to stress and shape their relation to an omnipotent being. This raised a critical question throughout my work: What does it mean to interview and write about a religious person? Even if one acknowledges a religious dimension in the person's life in the analysis, this dimension might disappear for the religious reader when such an acknowledgment is placed in an academic work like my own. Consequently, there is, as Rambo has argued (1993), often an inevitable gap between religious people's spiritual understanding of themselves and social scientists' writing of it. However, encounters across any kind of difference and how the understandings and lives of others are being represented are not new predicaments in anthropology. In this book, I attempt to offer, what I consider, a meaningful way to address this problem.

I believe it goes without saying that my intention is not to give a representative picture of conversion to Islam or an overview of female Muslim converts in Sweden or in the United States. Neither is it a comparative study between female Muslim converts in the United States and Sweden. Instead, the book is about some women in Sweden and the United States, their identity as Muslims and their personalized versions of Islam. Since the majority of the women in the study are Swedish, the Swedish context and situation are stressed slightly more than those of the United States.

Islam in the West

As mentioned, this is not a book about Islam per se. Nevertheless, a brief description of the main tenets of the religion and different discourses, as well as of the Islamic view of conversion, is apt. Islam manifests a belief that there is no other God than Allah, and that Mohammad is His prophet. The holy scripture of the Qur'an and the five pillars of the creed—*shahada* (proclaiming commitment to the faith), *salat* (prayer), *zakat* (almsgiving), *sawm* (fasting), and *hajj* (pilgrimage to Mecca)—as well as the *hadiths*, the traditions and examples of Mohammad,[23] and *Sharia*, the canonical law of Islam, provide the religious unifying framework for most Muslims.

Like other religions or ideologies, Islam takes different forms and expressions depending on social and cultural context. There are different trends on how the *hadiths* should be interpreted and understood (Roald 2001) as well as different general interpretations of Islam, often labeled as modernist, traditional, and fundamentalist. Moreover, there are modernist Muslim feminists as well as more secular feminists (El-Solh and Mabro 1994, Svensson 1996). The role of the Muslim woman has been seriously debated, in Muslim as well as non-Muslim societies, and different interpretations and representations have been offered. In one version, proposed by conservative Muslims, the role of the woman is confined to that of being a mother and wife, while modernists have used those reinterpretations of the same verses of the Qur'an as well as the *hadiths* that create a more "modern" understanding (Haddad 1998:12, Roald 2001, Svensson 1996). The veil is also a disputed and controversial issue. While some Muslims point to the Qur'anic verse that commands women "to cover their ornaments" and "draw their veils over their persons" as God's definite order to conceal their physical appearance, others among Muslims refer, as an argument against the veil, to the verse that is more specifically about the veiling of the wives of the Prophet Mohammad (Ahmed 1992:5, 55f.; El-Solh and Mabro 1994:8). We will see that the women approach these sensitive issues in different ways. The diverse interpretations invoke further problems and discussions regarding the conflicting arguments of socially and culturally adopted understandings of Islam, on one hand, and of a universal Islam, on the other hand.

A basic idea on conversion within the Islamic faith is that anybody can convert to and embrace the belief. This is ultimately based on a principle that all human beings are born as Muslims, and that they hopefully will find their way and revert to the only faith. There is no word in Arabic for "conversion," rather there is the idea of "becoming a Muslim," for which the verb *aslama* is used. *Aslama* means "to submit" and also it is from this verb that the words "Muslim," meaning "one who submits, a submitter" and "Islam," meaning "submitting, submission," are derived (Dutton 1999:151). Formally, the convert needs to make the formal declaration of commitment and faith, *shahada*, "There is no God except Allah and Mohammad is His Messenger," often recommended to be done in the witness of an imam and a second person. It is asserted to be a personal choice and action, not requiring any external approval. As one Muslim Web page expressed it:

> However, it would not be sufficient for anyone to only utter this testimony orally either in private or in public; but rather, he should believe in it by heart with a firm conviction and unshakeable faith. If one is truly sincere and complies with the teachings of Islam in all his life, he will find himself a new born person. This will move him to strive more and more to improve his character and draw nearer to perfection.

The light of the living faith will fill his heart until he becomes the embodiment of that faith.[24]

Many of the converts touched upon this idea of being a personal embodiment of the faith, following the examples and habits of the Prophet (the *Sunna*). The only type of proselytizing expressed and practiced by the converts themselves was being a living example in their everyday life, always being aware of what kind of impression people can have of them, conscious of how they act in public. [None of the women had been a target of Muslim missionaries but each had rather been witness to a Muslim way of life.] They encountered and learned about Islam through friends, colleagues at work, or through trips and stays in Muslim countries. They themselves had been the ones seeking out information (cf. Poston 1992).

Even if Islam, together with Christianity and Buddhism, is categorized as a missionary faith, there is a debate about its expressions and means, and only a minority seem to think that missionary activity is a religious duty for every Muslim (ibid.:3ff.). Interestingly, proselytizing or informing about Islam, *da'wa*, in the West seems to be directed at Muslim immigrants as much as non-Muslims. In many cases, Islam finds new support and revived interest from Muslims who have migrated to Europe and the United States, and Muslim organizations and Islamic centers find it important to inform Muslims about their faith and seize the opportunity to talk about the "universal Islam," the pure Islam, beyond the many different cultural interpretations and practices that the Muslim immigrants have bought with them to the West (Schmidt 2004:144f.). *Da'wa* in the West does not involve insistent strategies such as door-to-door informing but rather the formation of Muslim centers and organizations, offering seminars, talks, and services, publishing texts, as well as obtaining contacts and having dialogue with the surrounding society. The very idea of active missionary work, associated with Christianity, was expressively and strongly rejected by a couple of converts, a position supported by *Sura* 2:256 in the Qur'an stating, "There is no compulsion in religion." Instead the very personally chosen way to Allah by each individual is emphasized—as one convert expressed it, "there is said to be as many ways to God as there are breaths of creatures."

The Bigger Picture

The women's transformation and new identities reflect larger social and cultural transformations. Seen from afar and within a greater social context, their conversion to Islam points to changes often described through the phenomena of globalization and postmodernity. The presence of Islam in the West is largely an effect of migration, conversions, and the "flow" of ideas and thoughts. The improved and fast communication of the (post)modern world has also been a great tool for the networking and solidarity of the *ummah*, the worldwide Muslim community. What follows below is a brief guide to the two national and social contexts within which the converts live.

During the latter part of the twentieth century, Sweden, as a rather homogeneous and monoreligious Christian society, changed in significant ways due to immigration.[25] Most of the immigration to Sweden in the last two decades has come from countries dominated by a Muslim tradition, such as Iran, Iraq, Lebanon, and Ethiopia (Sander 1997:179ff.). Today, the building of mosques, the opening of Muslim schools, and veiled women in public are all a part of Swedish society. Given that there are no official statistics about the ethnic, cultural, and religious backgrounds of people living in Sweden, it is difficult to estimate the number of Muslims in this country. Instead, the available statistics are based on nationality, on citizenship, which is a deficient indicator of which religious tradition people belong to. The problem here is what definition do we give to being a "Muslim"— ethnic, cultural, religious, or political? (ibid.:184ff.). The estimations vary a great deal. The numbers mentioned by the press and the Islamic centers and organizations themselves range from 300,000 to 350,000 (Otterbeck 2000:109), while islamologists say that the number of Muslims who have been in contact with a Muslim community does not exceed the figure of 100,000 and that the number of those attending prayer each Friday is about 15,000.[26] Neither are there any official statistics about the number of converts and the same is true of the United States. However, I have got an estimation of 6,000 converts in Sweden from the converts themselves. Sander estimated two to three thousand converts over ten years ago (1991:87).

As mentioned earlier, Islam is today the fastest growing religion in Europe as well as in the United States. Being the second largest one in France and the third in Britain, Germany, and North America, it is of importance to consider the different aspects of Islam in the West. It was estimated that Islam will replace Judaism as the second largest religion in the United States in the beginning of the twenty-first century (Esposito 1998:3).[27] Muslims in America, as in Sweden or in any other country in the West, are far from homogenous, coming from different ethnic and national groups with divergent attitudes and practices. In the United States the majority of Muslims consists of first- and second-generation immigrants and African American converts. Recent studies assess that there are 2 to 7 million Muslims in the United States, composing 0.7 to 2.4 percent of the population. In 1992, the American Muslim Council, one of the largest Muslim lobbying organizations in the country, estimated in a survey that the main groups are African Americans (42 percent), South Asians (24.4 percent), and Arabs (12.4 percent), while American whites represent 2 percent (Schmidt 2004:2f.).

[handwritten margin note: muslims in the United States]

Even though the United States is a "country of immigrants" founded on religious freedom and multicultural ideals, Muslims, like other religious minorities, experience problems of integration, of practicing their faith in a society influenced mainly by Judeo-Christian or secular values, and of trying to continue their religious traditions while at the same time carving out a place in American politics and culture (Esposito 1998:4). The same is true of Swedish Muslims as well. Religious freedom allows converts to appropriate Islamic practices in their everyday life. However, the surrounding dominant culture often treats with hostility and discriminates against the converts' new faith and practices.

In his essay *On Toleration*, the political philosopher Michael Walzer (1997) discusses different types of systems of tolerance. Two of them, "nation state" and the "immigration society," are applicable and helpful for the present discussion. His description of the nation state corresponds in many ways to the Swedish society in which the "national culture" is strongly influenced by a relatively homogeneous majority culture. A central characteristic of the nation state is that its majority is rather "stable." This dominating group organizes the societal life on different levels in a way that reproduces and maintains the national culture and identity. This is partly reflected in Sweden's holidays, public symbols, and ceremonies, as well as in compulsory education. In chapter 5, we will follow the women's different experiences of encounters in their surrounding world. As we will see their insistence on being both Swedish and Muslim obviously questions and provokes debate over some public representations of what "Swedishness" supposedly is, representations manifesting ideas about cultural belonging and ethnic purity.

I will show how the women, through their conversion and veiling, advocate a multicultural Sweden in which difference is represented not only by immigrants but also by "native Swedes" converting to different worldviews and lifestyles. It has been pointed out that the dominant ideology of conformity in Sweden has to be challenged if we are going to achieve cultural diversity and express tolerance and respect for it. How pronounced the limits of recognition of cultural and religious differences is obvious when the women speak of their experience of a general resistance to their desires to wear the veil at work and in other public spaces. The common resistance shown not only toward women veiling but also, for example, toward the building of mosques questions whether we really do have freedom of religion in the practical sense of the word.

In comparison to the nation state, the United States as a "society of immigration," the other system of tolerance discussed by Walzer, is distinguished by the fact that the state is, at least in theory, neutral toward groups and views everyone as citizens rather than group members. For example, religious holidays such as the Christian Easter are not public holidays. This system is built upon the ideas of individual choices and that tolerance should not only be directed toward other groups but also toward differences within the group, implying that the members of a group should accept each other's different versions of faith. A central difference compared to the nation state is that the state of the immigrant society is regarded as not having a national identity, but only a political one. This is where the hyphenated identities come into the picture. The hyphen symbolizes the recognition of the ethnic background as well as a common American identity, some kind of Americanization of the defended difference. Walzer brings up an example: The hyphen that links Italian-American symbolizes that Americans recognize the "Italianness" as a cultural identity without political demands and that "American" is a political identity without strong and specific cultural claims. However, these are ideals on which the systems rest. Walzer mentions how Anglo-American ideas, history, and culture still dominate some aspects of America, such as the educational curriculum. On turning the focus on praxis instead, the everyday implementations of these ideals, things come in quite a different light.

Given U.S. foreign policy and a history of imperialism (Said 1994), Islam is frequently viewed and focused on as a political force and threat. The Iranian hostage affair in the 1970s, the Gulf War in the 1990s, and recently the 9/11 terrorist attacks and "the war against terrorism" in Afghanistan and in Iraq have colored notions about Islam and "the Muslim world" as a homogenous area and a "civilization," not least because of frequently distorted media coverage (cf. Said 1997).

Moreover, ideas from academia find their way out and are used on a political level. In an 1993 article titled "The Clash of Civilization," political scientist Samuel P. Huntington argues that world politics is entering a new phase in which the most central conflicts and antagonism will be cultural, between civilizations. This argument has quickly been picked up, and after the 9/11 terrorist attacks we have several times heard journalists, "the man on the street," as well as politicians and academics referring to the dichotomy of the West and Islam, and the conflicts between "the Christian" and "the Muslim" world, an idea undeniably voiced in the Muslim world also. This political tension and conflict shed some light on the general public feelings about the identity category of "American Muslim" in a larger context, maybe implying particularly those who have "converted" to it. The "Muslimness" is then viewed not only as a possible cultural and religious identity but also as one with political claims, threatening American political loyalties. From this perspective and with the background of a quivering patriotism, the veil becomes a waving, warning flag. Like any other categorization, the talk about civilizations and the clash between them (if it is even possible for civilizations to clash in the first place) is crude and inconsistent. To counter charged political statements about "us and them," the West versus the Muslim world, and "clashes between civilizations" and Islam as the new enemy, we need analyses that show nuances, contradictions, and complexities.

Both in Sweden and the Unites States Muslims are establishing and practicing their religion and faith through different institutions, forming a "Swedish Islam" and an "American Islam," respectively (Svanberg and Westerlund 1999, Schmidt 2004). A couple of the women that I interviewed in Sweden mentioned this phenomenon and identified themselves with a "Swedish Islam," pointing to environmental issues and social politics, ideas they find supported in Islamic discourse. As Svanberg and Westerlund put it, "Together with converts, this first generation of native Swedish Muslims participates in forming a Swedish Islam, molded in Swedish societal structure and against a Swedish sounding board, an Islam that could therefore be called blue and yellow Islam" (ibid.:10, my translation). Muslim converts have gained a significant role within the Muslim communities. They are often engaged in different groups and organizations, working together with "born-Muslims," engaging in important discussions about interpretation and practice of Islam in a Western context (Roald 2001). The converts move within and through many different cultural and social settings. These may be activities within Muslim organizations, labor unions, organizations for the protection of nature, nursing homes, schools and so on, creating different loyalties and belongings.

On the whole, individual life stories and the different personal versions and understandings within the same "category" become a valuable means to counterbalance the essentializing of cultures, or what Eriksen (1994) has called "cultural terrorism," as well as encourage the recognition of cultural and religious diversity, not only between and within "groups" but also between and within individuals. This book aims to be a part of that project.

About "Swedish" and "American"
Women—or What?

Even if strong criticism has been raised regarding older views of nation and culture as essential and stable entities, the category of "national identity" often appears to be the most natural and taken-for-granted point of reference for any kind of comparison, particularly when, as in this case, the material consists of interviews conducted in two different nations. But this study is not primarily about national identities or nationalities, nor do I have any intentions of making any comparison on the basis of "national cultures."[28] An immediate impulse one seems to have is to approach the women as primarily Swedish and American and then search for systematic differences. However, my analytical approach problematizes the premises of such assumptions. The material is also, evidently, too small to offer any legitimate conclusions or even hypothesis about likely similarities and differences based on the category of national belonging. I myself have struggled with the categorization of my material; if it is not primarily Swedish or American material—then what is it?

My focus and interest rest on some women's idiosyncratic versions of a Muslim identity and diverse external representations. I believe most of us would agree that we cannot talk about "the culture of X" as a bounded, monolithic category of which a certain group of people are a product (Strauss and Quinn 1997:6). Culture as well as public national representations are not simply copied into people's minds, an idea Strauss (1992a:9f.) has called the fax model of cultural internalization. National discourses or narratives cannot be equated with a common mentality called "Swedish." But national categories are handy and easy to employ; they are often used without question even within our own discipline, as if they have an essential and inescapable meaning. We risk here a reification of nationalities. Looking through the interview transcripts, I discovered my own frequent, sometimes unreflective, use of national categories in my communication with the women. Similarly, in one of my articles

(Mansson 2000) I used the folk category of nationality as an analytical tool, referring quite uncritically to "Swedishness" and "Swedish" values. We use them repeatedly in everyday life, but this is obviously no justification for applying a folk model of nationality to a research model. The difficulties of ignoring and questioning them are partly due to having grown up with them and to some degree learned them. But as Strauss and Quinn (1997:256) caution, "do not presume that your readings of public culture are the same as those of the people you are studying."

We are already acquainted with the problems of labeling persons by categories of identity such as race, religion, ethnicity, nation, and sexuality. They may not only limit our analysis but also misrepresent the persons who engage them, giving no room for the shifting, capricious self. Labeling the women as Swedish and American suggests that "being Swedish" or "being American" is a highly relevant aspect of their identity and that it has significant impact on how they understand themselves and the world. I believe that when the notion of "Swedish" or "American" identity is translated directly into experiences of individuals, there is a great deal of slippage (Linger 2003:11f).

I do refer to "Swedish" and "American" women but, as I suggest here, these are categorizations we should use more carefully instead of assuming beforehand cultural similarities or differences based on observable public representations and symbols. People make sense of themselves through different cognitive representations, using diverse categories and discourses, sometimes in rather unpredictable ways. The theory of cognitive models and the view of culture as "differentially distributed cultural understandings," offer a way to describe cultural sharing without reifying culture as a bounded entity (Strauss and Quinn 1994:293). This implies a view that cultural understandings are not distributed randomly. Rather, "culture" is the distribution of shared cognitive models (Wallace 1961, Schwartz 1978) or what Sperber (1996) calls "epidemiology of representations."

Disposition of Chapters

This study offers analyses on how the converts manage their identity formation internally (parts I and II) as well as externally (part III). Part I, *The Individual Conversion*, continues with chapter 2, "A Step toward the Unknown," where I offer a short biographical presentation of the nine women. It aims to give the reader a picture of them with respect to age, background, and how, where, and when they first encountered Islam. In chapter 3, "The Conversion Narrative," the interview material (the narratives of the women) is treated as an essential part of the conversion itself. Since it concerns the relation of narratives to experiences and the self, as well as the interview encounter, it can be read partly as an extended methodological discussion. It demonstrates the narratives as transitional phenomena, serving as important means to mediate and reconcile meanings between self and others, past and present, and "inner" and "outer." Narratives are analyzed as important tools to communicate a consistent self-presentation and offer a socially reasonable story as to why one converted.

In part II, *Personal Worlds*, I explore each of the women's subjective conversion experiences and identity formation in relation to personal models. Departing from their life stories I show how the women understand Islam as well as organize their identity through these mental structures linked to emotional memories. Chapter 4, "Personal Models of Spirituality and Social Conscience," focuses on the women who highlighted the spiritual dimension of their conversion as well as social issues, while chapter 5, "Personal Models of Gender," continues to elaborate on how the women reproduce as well as challenge dominant ideas, both Western and Islamic, on gender and gender relations. With a special focus on the personal meaning assigned to the veil, chapter 6, "The Veil and Alternative Femininities," discusses the veil as a personal symbol, which produces a critical feminist commentary as well as alternative femininities.

In part III, *Encounters*, the emphasis is on the women's interpersonal relationships and how these reflect the ways the women externally negotiate their Muslim identity. In chapter 7, "Looping Effects of Meaning," I focus on the tension between the women's self-understanding and the images and expectations of others, and the looping effects of meaning prompted by the women's active engagement with the surrounding world and various categories. Chapter 8, "Family, Work, Sisterhood," elaborates on the women's experiences of encounters with family, employers, and Muslim women and the strategies they employ against stereotypes and harassment. I conclude my analyses with some final reflections and conclusions in chapter 9, "Personal Versions of Islam."

Chapter Two

A Step toward the Unknown

A Biographical Overview of the Women

Women interviewed in Sweden	Year of birth	Year of official conversion
Ayşe	1955	1990
Zarah	1949	1984
Marianne	1965	1986
Lisa	1967	1994
Cecilia	1966	1998
Layla	1950	1976

Women interviewed in the United States		
Mariam	1947	1972
Fatimah	1954	1987
Hannah	1954	1971

Below I give the reader a brief biographical presentation of the women, presented in the same order as listed above. Their life stories will be discussed more thoroughly in part II, "Personal Worlds." There are obviously many different factors behind a conversion, conscious as well as unconscious, and though just a few might even be mentioned in the interview, most of the women expressed a longing and search for a change. They talk about travels to far-away places, experiences of other cultures that triggered reflections on their life, life crises, feelings of meaninglessness, and encounters with Muslim people. What the women have in common is that the step they took toward the unknown was just a first step in a very long process, often encompassing several years. I will begin with Ayşe. (Aye-sha)

Ayşe converted to Islam after many years of reading and questioning, as well as two visits to her husband's home country in Africa. Her first trip made a major impression on her as a young teenager.

She was then an atheist, an outspoken socialist, who had protested and participated in manifestations against the U.S. involvement in Vietnam and for solidarity with the "Third World." Her husband was born and raised in a Muslim family but he was not, and is still not, a practicing Muslim. Just like Ayşe, he too was an atheist and a convinced socialist when they met in the 1970s. Ayşe was born in the mid-1950s and her parents were secular, seldom talking about religious issues and matters. She was confirmed, but it did not really have any religious meaning to her. Later, she left the Swedish church. She was the first one in her family to study at the university and she took classes in both Russian and Arabic.

Ayşe was very much taken by her experiences in Africa. She was fascinated by what to her at that time seemed exotic cultural traditions, music, and dance. It was not until a decade later that she came into contact with a Muslim women's organization, and then it took five more years until she converted officially. She had, however, lived as a Muslim for a while before. At the time of the interviews Ayşe and her husband had three children. She was working at a center for Muslim children and was involved in various women's organizations, visiting classes, and helping Muslim immigrant women in their everyday life in Sweden within different projects.

Zarah likewise left the Swedish church at the age of twenty-one. As a child, however, she attended Sunday school and had been confirmed. She had a belief in God but she wanted to feel "free" from the church and the ideas she had grown up with, which she had difficulties grasping and accepting. Zarah felt a strong aversion toward the negative images of people of other cultural and religious backgrounds, images that were reflected in the missionary idea of the white man's mission "to convert the heathens in Africa." During her time in college she did some traveling, visiting Turkey and Germany where she met her husband-to-be, who was born a Muslim. He came to Sweden and she married him at the age of twenty-two. Her husband was not a strict Muslim and did not have any particular wish that she should convert to Islam. At that time she was not very interested in Islam, but seeing it practiced in her everyday life, not only by her husband but also by their friends, triggered her curiosity. Zarah and her husband also traveled a couple of times to his home country, and these visits made a strong impression on her. Not until thirteen years after their marriage did Zarah officially convert to Islam. During this long period of time she had read about the religion, talked to Muslim friends, and obtained a better understanding of it and how it was lived. Shortly thereafter she and her husband were divorced. She continued to read more and she also became active in a Muslim women's group. Several years later she met another man to whom she was married for a year. At the time of the interview she was working with children at a daycare center.

Marianne was born in the mid-1960s in a family of five. She was not raised within any religious tradition, quite the opposite. She was an atheist, she was not confirmed, and she considered everything to do with the Swedish church to be hypocrisy. Marianne came into contact with Islam through a close friend, and what started out as a strong negative reaction, doubts, skepticism, and questioning slowly turned into a belief. She converted officially in her early twenties, a year after she had married her Muslim husband. In response to pre-conceived notions that Swedish women convert to Islam because of a man, she pointed out that she married him knowing that she wanted to become a Muslim. They married shortly after they met, according to Muslim tradition. Like some of the other women, she is very active in different women's organizations and is engaged in fighting for women's rights. Marianne has gone through an extended process of reformulating a female Muslim identity. Today she and her husband have four children and she works as a teacher at a college.

Lisa is the youngest of the women I have met, born in 1967. Her parents are not religious, and she did not really think about religious issues herself until she went through a life crisis in her mid twenties. She had ended a relationship with a man that was not good for her, and in relation to this she felt a stronger need to have some connection to God. At that time she was working as a teacher at a school, where she got to know some Muslim people from whom she borrowed books about Islam. Lisa read more and more and after some months she went to Turkey. There she was, for the first time, accepted as a Muslim. She described the time as a "gray zone," not knowing exactly what she was, but the visit was important to her. Compared to some of the other women, Lisa was rather alone during this time. She did not have any support from a Muslim group or friends; it was a lonely period of praying and reading. Today she is married to a Muslim man and works as a teacher for immigrants.

Cecilia is also one of the younger women. She was born in 1966 in a medium-sized town. Until the age of eighteen, she was active in the Swedish Catholic organization for youth. After senior high school she traveled around for several years. First she went to Israel to work on a kibbutz, and what was supposed to be an eight-month visit turned into a three-year stay. When Cecilia came back she worked at a hotel for a couple of years and then, in her mid twenties, she went abroad again. She worked as a receptionist and as the captain's secretary on different cruise liners in South America, in the Caribbean, and the Mediterranean. After some years of living on cruisers she returned home. Cecilia was then twenty-eight years old. When I met her for the first time in 1998 she was working at a nursing home. Cecilia has also taken different classes at the university in East Asian culture, political science, and English. She felt that she had had enough of a life moving back and forth. At her work there were some religious Muslims and from them she started to borrow several books about Islam. She became more and more intrigued and it all seemed to make sense to her. It took five years of studying and reading before she took the step to convert. When I met Cecilia she was newly married to a Muslim man. She had met him through friends in common, and two weeks later they married. They had met a couple of times before, always with a witness attending according to Muslim tradition. At the time of the interviews, Cecilia was pregnant with her first child. Her plans were to continue her university studies after being home with the baby.

Layla, born in 1950, was not from any particularly religious family. However, as she stressed, she has always had a belief in God from which she has never departed. In her mid-twenties she traveled with the "pink buses" for six months to India. During the same period of time, she met a Muslim man at her work that caught her attention. When Layla came back from the trip they got to know each other better and in 1976 they married. He was not a Muslim, in any spiritual way, as she expressed it, but he did not eat pork or drink alcohol. She formally converted to Islam before the marriage, but she went through the ceremony without any religious conviction or knowledge about the religion. It was only some years later that Layla started to read more seriously. She also met other Muslims and people who were in the process of converting and with whom she discussed the religion. Like most of the other women, Layla went through a long period of testing and thinking. Later on she became active, going to lectures and traveling to international conferences. It was not until 1990 that she started to wear the veil. Layla too is working with children. At the time of the last interview she and her husband had four children.

Mariam was twenty-five years old when she converted to Islam "in the field." She was a young graduate student of anthropology doing her fieldwork in a little oasis in northern Africa. She came back to the United States three years later as a Muslim. Mariam was born on the East Coast of the United States in the late 1940s. She was the only child in an upper-middle-class family, with a mother of Jewish background and a much older father, who was Catholic. Her mother had converted to Christianity and was all her life, in Mariam's own words, a "searcher," looking for religious meaning. Mariam was brought up in a spiritual and intellectual milieu; both her parents were personally interested in religious issues, and they had decided to give Mariam a Catholic upbringing. She studied subjects such as Greek and Latin, archaeology and anthropology, and during the summers in college she did fieldwork in Europe and Africa. An early fascination for northern Africa and the Sahara made her choose as her field of study a small oasis in the middle of the desert where some "seminomadic" people lived. As a curious and adventurous young woman in her early twenties Mariam arrived there in the midst of the ongoing war between Libya and Chad. She was "sent into a very military sensitive area," an area that was considered very remote and dangerous. The people she lived with for three years welcomed her, and living among them she learned their way of life. These years stand out as a very happy and extraordinary time in her life. It was during Ramadan that Mariam experienced something she could neither ignore nor explain with any analytical arguments.

Today, she and her husband belong to a Sufi group in California. Her husband, who was born in a Jewish family, had also converted to Islam (more specifically, to Sufism).[1] Mariam met him when she came back to the United States after her fieldwork. Today they have five children and belong to a small Sufi community. Mariam does not work outside the home but runs a "home-school" for two of her children.

Fatimah belongs to the same Sufi group as Mariam. She was born in the mid-1950s, in a city "in the middle of Western America," and, like Mariam, was brought up in a home where religion was an important part of life. Fatimah was raised Catholic as the fifth of eight children. She liked the stories of the saints and still remembers the feeling she had as a child of being uplifted when attending the mass. But as she grew older and entered adolescence, she had difficulties with what she experienced as paradoxes in the Christian faith. When she started college she threw away everything that had to do with religion. A year after graduating, Fatimah met a man whom she married shortly thereafter. She was then in her early twenties. They had two children and moved to New Mexico where they started a business together. It was about five years later that she experienced a deep existential fear when watching a documentary about nuclear holocaust. This was the beginning of a spiritual quest, a period of reflection, reading, and trying out prayers. Fatimah came into contact with a religious, ecumenical center, where she had some crucial spiritual experiences. After a couple of years she divorced her husband and returned to the center, where she converted officially. Some years later, through the center, she met a Muslim man, he too an American convert, and they married. Today they have two children and run a family business.

Hannah was born in the mid-1950s in Mississippi but grew up in the Midwest. She was raised in a Christian tradition and frequently went to church on Sundays. All her relatives are Christians and her mother's father was a minister. After high school she came into contact with some Muslim people and felt drawn to what she learned about Islam and the way they were living the religion. Of the several elements of Islam, Hannah was particularly attracted to the emphasis on family and the different roles of men and women. She started to read by herself and soon thereafter she met her husband-to-be, who had converted to Islam years before. They married at an early stage of their acquaintance according to Muslim tradition. She was then seventeen and he was twenty-three. A year later Hannah took the *shahada*, the formal conversion ceremony.[2] As an African American she had paid some attention to the Nation of Islam but she did not feel compelled by their focus on black history and how it exclusively turned to African Americans. "In a religion, in Islam, everybody is like sisters and brothers," as she put it. She and her husband have five children and have also become grandparents. Hannah works full-time at a children's center that is a part of a culturally and religiously diverse college and in the evenings she takes Islamic classes.

Chapter Three

The Conversion Narrative

Conversion is often viewed as a sudden and fundamental shift in worldview, which changes the individual in considerable ways. William James wrote in 1906 that to be converted signifies the change "by which a self hitherto divided, and consciously wrong, inferior and unhappy, becomes unified and consciously right, superior and happy" (Harding 2000:33). In the classic study *The Social Construction of Reality*, Berger and Luckmann (1984 [1966]) describe the alternation of religious conversion and the transformation of subjective reality. A successful alternation requires, among other things, that the old reality be reinterpreted within the apparatus of the new reality. They argue that this reinterpretation results in a rupture in the subjective biography of the individual in terms of "B.C." and "A.D.," that is, pre- and postconversion. Everything in life before the conversion is now understood as leading toward it, and everything following it as flowing from its new reality. Formulations such as "Then I thought . . . now I know," which are common in the conversion narratives, reflect this kind of reinterpretation of earlier experiences and actions (ibid.:179). However, for the women in this study there was no "biographical rupture" in the sense of what Berger and Luckmann call a "cognitive separation of darkness and light" (ibid.:180). Naturally, the women reject certain values and ways of living, but they seldom express a fundamental denial of their previous life or self. Instead, the women reorganize their biography and through a conversion narrative they create self-coherence and continuity by negotiating meaning between past and present, between the one I was then and the one I am today.

As I discussed in the first chapter, I understand and treat conversion as a continuing process rather than a one-time change and a total break between a "before" and an "after." This is not only a theoretical assumption, it is a hypothesis that rests on the converts' experiences. In comparison to much of the experiences analyzed within the conversion literature (which in the West has mostly focused on the Christian tradition), most of the converts highlighted the official conversion not as a particular sudden moment of intense emotion but rather as the outcome of a long period of reading, talking with Muslim friends and colleagues at work, as well as visiting and living in Muslim countries (cf. Poston 1992). Also, beyond the particular moment of the formal conversion ceremony, which is not particularly stressed by the women, their narratives of becoming Muslim reflect changes, but while talking about these life-changing experiences they also mediate coherence in life, feelings of "being back home." In the midst of transformation the new belief is understood and integrated within a particular subjective, interpretative world. Becoming Muslim presumes not only a change in the convert's internal world and actions but also in the shaping and telling of her life story. A main idea underlying the discussion in this chapter is that the conversion is a continuous process of integrating a rupture in worldview with a coherent life story, an integration that is achieved here during the interview exchanges, through the conversion narrative. Thus, the conversion narrative, the talking about the conversion, constitutes a significant part of the conversion itself. The conversion *evokes* a conversion narrative. I attempt to show that the narratives are strongly motivated, prompted by feelings and reflections, intended to assign personal meaning to the changes in life and to mediate a sense of continuity and a certain self-understanding to others.

It is conspicuous that the act of narration is an important psychological and social tool to demonstrate a consistent biographical account and to come to terms with and reflect on particular events and experiences. The presentation of a coherent self, between the one I was in 1976 and the one I am today, as well as being able to give socially acceptable and comprehensive explanations are fundamental functions of the narrative. The conversion narrative is thus a salient means through which the women negotiate their Muslim identity and try out alternative ways of making sense of the transformation in life. The event of narration not only organizes events and experiences into a tangible and coherent biography; it also makes possible the integration of a religious system into a personal life story (cf. Stromberg 1985). To tell a conversion narrative is thus not just to represent the transformation but also to give it meaning as well as to relive and strengthen it in the moment of telling. With narrative I mean here story or talk. Rather than approaching the narratives of the women as all set, completed "texts," I distinguish their transitional character reflecting an ongoing meaning-making in the particular encounter of the interview.

Narratives as Transitional Phenomena

As an anthropologist studying cognitive/emotional and cultural aspects of identity and conversion, making analysis and assumptions based on interviews, I have to rely on what people say. I thus depend on linguistic information—but this does not imply that I disregard the existence of nonlinguistic knowledge. For example, as I will show in the next chapter, in the case of Mariam, the initial learning and understanding of Islam happened through the very *practicing* rather than reading and talking.

Bloch (1998), a cognitive anthropologist, argues that much of our knowledge is nonlinguistic. From his twenty-year fieldwork among the Zafimaniry in Madagascar, he discusses mental models that are not simply linguistic, "but partly visual, partly sensual, partly linked to performance" (p. 26). Similarly in the tradition of phenomenology it has been stressed that meaning also lies in the doing and therefore should not be reduced only to what can be thought or said (Jackson 1996:32).[1] As Bloch points out, it has become common among anthropologists to view the world as constructed through narratives and texts and to say therefore that there is nothing beyond the reality created in these, since any other past or present is invisible. He continues, "What I find totally unacceptable is the notion that cognition of time and other fundamental categories is constructed through narratives and that consequently an examination of narratives will reveal directly a particular group of people's concepts of the world they inhabit" (1998:102).

This implies two important statements, with which I agree. First of all, the narratives do not construct the self-image of the convert. *Individuals* construct a multiplicity of narratives in interaction with others, in different contexts. Secondly, the individual's knowledge and cognition is not bounded by one single narrative. Therefore, we cannot assume either that we could possibly have access to a person's whole subjective world and its complexities or, in other words, that the narrative can be equated with the world of the convert (ibid.:110). Consequently, I treat the narratives as neither self-construction nor self-revelation. Instead, they suggest subjective reflections, dialogues, and social interactions, in which meaning is negotiated and constructed between participating selves.[2]

"Talk" is, however, as Quinn and Holland stress, one of the most important ways in which people negotiate meaning and achieve social ends (1987:9). The interview offers an opportunity to explore different ways to successfully mediate the personal meaning of conversion, not only for me but also for themselves. Linger, who shares substantial portions of conversations from his fieldwork with his readers, has fruitfully referred to the interviews as experiments, due to "the interviewees' explorations, self-questioning, and dialogic play during our conversations" (2001:310).

Consequently, narratives are something else than mere linguistic expressions. Narratives, and this counts for language too, become meaningful when it can convey the personal experiences and feelings of the narrator satisfactorily. This act might even imply a struggle when the experience is not necessarily linguistic in the first place. Neither the verbal interaction nor the conversion narrative constructs the convert's sense of self, but they do reflect the individual's serious attempts to render personal emotions and experiences by employing available categories and representations in idiosyncratic ways.

Compelled by the desire of self-exploration and to communicate the personal motives of the conversion to others, the narratives have both an inner and outer dimension. The conversion narratives are *transitional phenomena* in so far as they communicate, mediate, and create both personal and cultural meaning. As a significant part of the conversion itself, they are important means in "translating" personal matters into culturally and socially recognizable forms and ideas, mediating continuity and change, and reconciling the psychological unease caused by the gap between one's inner sense of self and the socially imposed categorizations. I treat the narratives as transitional in three respects—they negotiate and transfer meaning between self and other (them and me), past and present, and "inner" and "outer." This chapter is organized around these overlapping themes.

The Intersubjective Encounter

We are constantly immersed in dialogue with others and with ourselves through introspection. Besides the specific dialogue between the convert and me, several other dialogues are manifested during the interview with the women, such as dialogue with family, with friends, with "society," and with God, and also with the one she was before.

However, each narrative took a particular shape depending upon the two participants, the convert and I. I have mentioned how my presence and identity, being a young non-Muslim female, and their perception of me inevitably shape the dialogue and interaction. It is easy to refer the displayed reality of the interview only to the world of the informant, but similar representations are misleading. The narratives acquire meaning and are interpreted from the subjectivity of the researcher, the subjectivity of the informant and the "intersubjective encounter" that they create together (Chodorow 1999:266). The ethnographic knowledge is then a part of an "ethnographic third" (ibid.:213), a potential space that takes shape during a mutual construction of reality (Crapanzano 1980, Hastrup 1992). The dialogue between the women and myself is composed of the relationship between their description or utterances and my responses to these. The interview exchanges seem to operate in a third space, the in-between space between self and other. The women's accounts of themselves and events and the interpretations of these are directed not only at me but also at themselves as well as to imaginable others. During the interview the women try to communicate a mass of perceptions, feelings, and ideas by engaging diverse belief systems and public representations. Through the narrative a certain meaning is created and negotiated between the convert and myself.

Throughout the interview different self-representations were tried out. Not only did the women negotiate different selves but in the interaction different self-representations of me too were manifested—as an interested researcher, a sympathetic listener, a potential biographer representing their conversion story, a sounding board for them, someone that seems to understand or at least tries to understand, or maybe even as one of the "others" (cf. Collins 1998). (These are the selves that at least I think they might have experienced of me.) I was not a passive observer or listener but someone who, by telling certain things about myself, wondering, posing questions, reconstituted myself in the encounter with the women and encouraged them to develop certain themes and ideas. Narration is then not only an attempted retelling of previous experiences but also something that *does* something to the participants; it generates thoughts and new experiences, it is a part of an ongoing identity-making, not only for the interviewee but also for the interviewer.

There are cultural models for how a life should be depicted, and some important events are emphasized to convey a comprehensive picture of the path of the individual's life. In my encounter with the women, I made clear my purpose and interests. Their conversion is my objective, and this explicit and conscious focus unavoidably had an impact on the form of the narrative. I started to ask them about their background, when and where they were born, their childhood and adolescence, and hence formulated a kind of life story that also functions as a chronological frame. I was given a conversion narrative where the conversion as a turning point was put in focus, and from which the women selected certain events and experiences. Even if they most often followed a temporal order, the direction of the narratives is dependent also on the women's attention to moral issues as well as on their relationships with family and friends. In this regard, the women's accounts are quite value-oriented.

To narrate and to "have" a story is understood as something essential in order to know who you are and also to be able to communicate this self-image to the surrounding world. However, it is a cultural idea that every normal functioning person should have a life story (Linde 1993:20) and that if they do not, their actions may seem less rational or thought out. In most cases the women offered well-organized versions of their life story and were, in my view, rather successful in conveying a coherent account. Zarah was, however, one exception. I will let parts of our conversation serve as examples of how the narrative is created within a particular intersubjective encounter, reflecting the negotiation of meaning between self and other. As the reader will notice, the interview was maybe not very "successful," ethnographically speaking, but nevertheless interesting.

In my encounter with Zarah, the interview itself became a moment when Zarah openly and critically contemplated the different reasons why she actually converted to Islam. In the dialogue below, she is stepping away, looking at her own life from the outside, as the other, asking herself as much as trying to give me answers. After each of my questions and her own responses to these, there were often long silences.

> *Anna:* After a while you felt "why not [convert]?" what do you think made you think like that from previously thinking "no, I will never convert"?
>
> *Zarah:* Before I thought Islam didn't fit with a modern society.
>
> *Anna:* What made you think more positively about it [Islam]?
>
> *Zarah:* How could that be? Could it possibly have been because I had experienced so much friendliness among people?
>
> *Anna:* That you had got more and more experience—
>
> *Zarah:* Yes, I believe it was. I felt safe with it as a human being . . . not all [of her and her husband's friends] were Muslims practicing the religion. What could it have been?
>
> *Anna:* Did you see more Muslims than Swedes?
>
> *Zarah:* No, I didn't do that either. As a matter of fact I didn't.
>
> *Anna:* I'm just wondering.
>
> *Zarah:* Now when I think of it . . . the ones we did see were very secular.
>
> *Anna:* Do you think you searched for something because you wanted to believe . . . do you understand what I mean? That one is driven by the will to do it? I don't know.
>
> *Zarah:* Well, I wonder. I do not remember. That is why I'm thinking . . . I remember that I have felt a longing, have been searching . . . [unclear]. I have simply only lived here. [Unclear] I have not given much thought to it.

Anna: Still there is something that has changed your attitude.

Zarah: Yes, there is. [Unclear] I cannot point to anything.

Anna: There are no simple answers to the question "why did you convert to Islam?"

Zarah: No.

Anna: There are so many different factors, it is hard to explain.

Zarah: Yes, it surely is.

Anna: It is probably hard to explain what you were thinking *then*.

Zarah: I surely would need to scrutinize myself and go through and try to find . . . that would be interesting. It would be interesting to see how thoughts and attitudes have changed.

Anna: You started to read more.

Zarah: Yes, but I cannot say either that those were the words that . . . I have met people who have told me that "it was when I heard somebody say that at a seminar" and that was the part of the puzzle. But I have nothing to point at.

Anna: No, you don't need to.

Zarah: It is so different depending on how we are. I do believe in God and I think that he knows what is best for me. That is why I have been pushed this way then.

After exploring certain ideas and explanations, asking herself "what could it have been?" she concludes by designating a religious explanation. While some women made a full performance during the interview, offering a more or less well thought out version of their story, Zarah seemed to search for, in regard to our specific situation and interaction with me as a researcher, seemingly intelligible explanations. Unlike the other converts, I only met Zarah once, which is why it is hard to analyze why the interview differs in this regard. She was also the only one that did not want to be interviewed a second time.

Some of the women placed themselves in a position of authority while narrating. Lisa, Marianne, and Ayşe are all in the habit of telling their story when visiting schools and giving talks. Zarah, on the other hand, is probably less accustomed to situations like these and is therefore uncomfortable. Besides not having a more "readymade" story, it could also be an unwillingness to share and reveal personal aspects of her life to a stranger. I believe she perceived the interview more as a questioning than as an opportunity or occasion to tell her life story, though all along I tried to create an informal atmosphere.

The dialogue between Zarah and myself was often filled with long silences. When reading the transcript I see how I offered suggestions trying to lead her into different possible thoughts and explanations. This brought awkwardness into the conversation. I felt at times, both during the interview and when rereading the transcripts, that I was pushing her for answers and that she may have felt uneasy and stressed for not being able to give concrete answers and to verbalize certain ideas. After one long silence she commented: "The tape is running without . . . I do not know what to say. You have to ask me more questions." At one point I asked her about encounters between "Swedish" and "Muslim" values and in what ways some of her previous ideas might had changed or been questioned. She found it difficult to come up with some examples, but later in the interview she told me about a situation in her children's school.

> *Anna:* That is an example of how a Muslim idea has changed your way of looking.
> *Zarah:* Yes that's right. *Finally we find something* [my emphasis].
> *Anna:* Well, I might be unclear.

Unlike the other women I met, Zarah did not seem to see herself as being in the center during our encounter. Instead it was our interaction and the encounter that were much more in focus; "we" finally came up with some possible answers. Since I felt frustration on her part and was worried that she would feel troubled and uncomfortable, I commented that my questions were rather extensive and maybe vague. In the end of the interview Zarah reflected on what was bothersome and unsettling for her.

> *Anna:* Is there something you feel that I have forgotten to ask you about, that you find important?
> *Zarah:* I have such difficulties in finding examples . . . it is hard to take away the Muslim side and look at oneself.
> *Anna:* Is that the way you feel?
> *Zarah:* Yes, that is what I have to do—
> *Anna:* —to get perspective, distance?
> *Zarah:* Yes, because I feel like an ordinary Swede in first hand . . . but then when you look I'm not that . . . but to bring out the specifics one has to see it at a distance.

Our conversation seems to have elicited critical self-reflection. Above, she brings up a crucial issue regarding narrating, particularly about one's own experiences and feelings. In dialogue with me Zarah is trying to give a self-presentation and to elucidate the conversion and her Muslimness but at the same time points out how difficult this is to her. It means, as she so rightly puts it, to "look at oneself" beyond the Muslim identity and "to see it at a distance," questioning and exploring the taken-for-granted aspects of one's self and signifying these from an outside perspective, that is, from the other's viewpoint. She is a Swedish Muslim, which for her entails a feeling of being "an ordinary Swede."

Something else should be added. This interview was one of the first ones I did at an early stage of my fieldwork, at a time when I somewhat thought of my project in terms of "difference" and "national categories." It might have been hard for Zarah to relate to my categories and questions, which did not necessarily have any relevance to her own experiences. She does not give as coherent a conversion narrative as the other women, but by referring to her belief in God and to a religious assumption ("I think he knows what is best for me") she offers some explanations. She has been "pushed" by God in a certain direction to where she is today. In the encounter between Zarah and me, at least her Muslim identity and belief had been confirmed.

Past and Present

A sense of continuity helps link elements of psychic reality and the sense of self into an alive "I." The power of feelings comes to be couched in terms of an individual's own history, but a history interpreted, absorbed, and actively created. It is recognized as one's own. Current feelings, a contemporary sense of self, passions, and felt needs and desires come not only from what really happened in the past but from a web of internal processes that construct the present. (Chodorow 1999:271)

The internal processes that construct a contemporary and coherent self are reflected in the narrative. The convert draws on the past not only to negotiate a sense of continuity but also to display transformation. But as Chodorow argues, the past is not merely what happened then but also the meaning that is given to the past in the present. The conversion narrative is motivated here in the sense that it serves as a means to explore and contemplate who one is today in relation to who one was earlier. While narrating, the women relate to and distinguish the past and incorporate it into the present. The past makes sense of existing feelings and vice versa.

Since it is the conversion and the women's identity as Muslim that are the reasons for the interview, it never took long until our conversation reached the moment of their encounter with Islam. The women who had a religious background and spiritual experiences when young stress this and connect it to their conversion and Muslim identity. For example, Layla emphasizes immediately her belief in God as a child. By connecting to a religious background she demonstrates temporal coherence and that conversion is a natural step that is sometimes even a fated decision.

> *Anna:* Where were you born and how old are you?
> *Layla:* I was born in the middle of Sweden. I'm 48 years old. Hmm . . .
> *Anna:* Can you tell me about your upbringing?
> *Layla:* Hmmm, I do not know what to say. I went to Sunday school when I was small. I have always had a faith in God, which I have never left.

Layla's "faith in God" has significant meaning, linking past and present, constructing a sense of life continuity. Similarly, Fatimah, who was also brought up in a religious home, initiates the interview by telling about her religious upbringing:

> I was born in Kansas, in the middle of Western America. And I was raised Roman Catholic and I was the middle child. I'm the fifth in a line of eight kids so we were a big family. And Catholicism was always really an important thing in our family. Religion was.

Her narrative then quickly moves to the turning point, the crucial event. In between, she incidentally mentions other important events in life, such as her trip to Europe, giving up religion in college, her marriage, and the business she ran with her husband. They are mentioned in a temporal order but are given less attention since she does not perceive them as essential in her explanation of the conversion. They are not understood as having any indispensable part in her conversion narrative. After a few minutes of the interview she reaches the culminating event of watching a documentary that she says, "hits this real core in me of fear about everything." Through the essential quest for spiritual meaning, Fatimah organizes self-continuity in her biographical account, from her childhood until today.

However, by referring to the past the women stress not only continuity but also self-transformation. In some cases the women relate to the one they were before the conversion almost as an other, one's previous self. The narrating "I" distinguishes the point between "now" and "then," and "here" and "there." The perspective of the narrator embodies a "here" and "now" while the other is understood to be in a "there" and "then." Linde elucidates this as the discrepancy between the narrator and the protagonist in the narrative. The separation between these is created by our self-reflexivity, our ability to relate ourselves externally, as the other. Within the dialogue between the self and the other taking place through the narrative, the use of values to specify the differences becomes significant (cf. Linde 1993). The converts distinguish themselves from who they were "then and there" before becoming Muslims, by stressing the difference in moral values and lifestyle. They view themselves in light of the conversion and its meaning and consequences. Fatimah gave a lucid example of this kind of dialogue between the narrator and the protagonist during our conversation, drawing on the difference between past and present.

> Thirteen, fourteen years ago I used to run a lot and I used to wear what other runners wear. Tank tops and shorts while running. Now when I look at that, I see it so differently when I see women doing the same things as I did. I do not judge them thinking they are bad people because they are doing it, but I see it in such a different light today. I really see it as a way of exuding sexuality, and tempting men. Whereas I really rebelled against anybody who said that to me fifteen years ago . . . But you know it is like stepping back from something and looking at it . . . I look at it in a very different light.

Fatimah "here and now" contemplates and evaluates the earlier Fatimah, the one she was before the conversion. In a retrospective light she makes a moral revaluation of her previous life, of the other Fatimah, from her new point of view. While narrating the convert takes a standpoint on the conversion, making a subjective evaluation of it and her life. Fatimah, the "I" of the narrative, observes her previous self simultaneously as this self operates as the other in the reflection upon Fatimah today and the changes she has gone through. She describes an act she previously performed without any thought, an act that she today regards as wrong and immoral. In the self-contemplation, her Muslim identity as well as her way of living become confirmed. Moreover, the Muslim belief system, here a particular religious view of covering women as "right" and "good," becomes integrated in the narrative. The moral changes that the spiritual experiences have brought with them need to be incorporated into the personal life story. The meaning Fatimah gives Islam is accentuated and her personal life and the Muslim ideology become intertwined (Stromberg 1985).

Telling our life story enables us to reflect on and evaluate the past from the perspective of today. Through the narrative the convert is able to observe and revise the self that is presented and formed. These are Ayşe's reflections:

> *Ayşe:* I really wonder if I hadn't become Muslim—
> *Anna:* —what would have happened?
> *Ayşe:* Yes, one would probably just go on trudging in the same track, one's whole life. And not developed . . . Before you lived a more normal life with work, the children, and being at home, meeting friends and that on the weekends. But now you are so active, thanks to Islam, in different organizations, visits to schools and I do not know what. It has taken a lot of time. And it has taken time from the family, all these engagements. Maybe it has become a little too much. I've almost got burned out.

Ayşe constructs a hypothetical past, illustrated as gray and dull in comparison to her busy, stimulating, and colorful life as a Muslim. In the converts' narratives we can locate a dynamic tension between continuity and transformation, between past and future, a tension that exists in the moment of telling.

> *Anna:* Do you still have your Swedish name, or?
> *Ayşe:* Malin Ayşe, yes, but mostly Ayşe.
> *Anna:* You do not use Malin?
> *Ayşe:* No, it is there on the papers and so [laughs].
> *Anna:* Your name has a lot to do with identity.
> *Ayşe:* Hmmm. Nevertheless, that is my name . . . Malin . . . my parents still like to call me Malin. And that is fine. Even the relatives in my husband's home country call me Malin since that was my name the first time I came there. Also they still call me Malin even if my name is Ayşe. They can call me whatever, it does not matter.
> *Anna:* When you think of Malin, do you think it is two different—
> *Ayşe:* No, I do not. I'm still Malin . . . you are after all the same person. I feel Swedish in the first hand. I am Swedish. It is not at all a transformation like that, so to speak.

In this rather complex comment, the meaning of then and now, of continuity and change of self, is negotiated through the use and self-experience of her two names, her given name Malin and the taken Muslim name Ayşe. In the conversion narratives two concurrent processes of self-making are expressed. Ayşe is still Malin. Besides stressing the changes and new possibilities, it is of uttermost importance to stress that one is still the "same person," a consistency between Malin before the conversion and Ayşe thereafter.

"Inner" and "Outer"

We experience flashes, textures, smells, pressures, and ghosts of emotion that cannot be languaged . . . Our own internal experience, if we permit ourselves to notice this, is a self without armor—perhaps without boundaries as definite as we would like—walking around in a world of others who appear to have proper boundaries and effective armor. Hence we may perceive ourselves to be in an alarmingly vulnerable position that must be remedied. And the remedy is to narrate, to create a self as other, replicating our experience of the actual others we seem to experience. (Linde 1993:121)

In the above quote, Linde expresses a sense of difficulty and emergency of rendering experiential and emotional stuff into meaningful language, translating personal meaning into ideas that are shared and recognizable to others. The conversion narratives reflect an ongoing, dynamic mediation between "inner" interpretations and personal experiences and "outer" structures, ideologies, and conventions. Through meaningful cultural themes the women attempt to socially express and mediate lived experiences in their narratives.

Through certain shared meaning systems, culturally familiar themes that the women can identify with and recognize as their own, they try to put words to their experience and make it comprehensible to the world. This section deals with how the women present a certain conversion narrative that translates "inner" to "outer," drawing partly on cultural representations that are used to objectify personal experiences. The converts often seem to use moral discourses, claiming how life should be lived. These can be ideas about gender roles, political standpoints, or religious ideals.

Narratives can be understood and read on several different levels. In part II, I explore the psychological salience of the ideas expressed by the women, and how experiences and knowledge are organized within a cognitive framework. Here I demonstrate the converts' attempt to create a coherent biographical account as well as offer an explanation, maybe not acceptable to everybody, but at least identifiable, as to why they became Muslims by employing different shared meaning systems. Through the use of certain sets of ideas the women seem to culturally and socially mediate their self-understanding. In a similar vein, Linde (1993) shows how cultural understandings, or what she calls "coherence systems," are invoked in life stories to rationalize behavior and make sense of one's life and action.[3] Below, I give two examples departing mostly from the narratives of Ayşe and

Fatimah. These accounts also reflect and serve as examples for the transitional meaning between past and present.

A Social Meaning System

Let me start with an excerpt from the interview with Ayşe to show the reinterpretation of past events and experiences through a particular meaning system and how a "turning point" is reconstructed in her narrative. For her, the visit to her husband's Muslim home country composed a triggering experience.

> *Ayşe:* It was a strong experience but maybe I didn't really understand or couldn't formulate it right then. But I have understood that afterwards. Then I started to think a little and some time passed after the trip, almost eight years, before I came into contact with the Muslim women's organization. They had classes on the Qur'an. Then I started to go there and I met other women who were on their way to converting. Then at least five more years passed, and I read more and checked things up. Because I was very hesitant about Islam. It is, to be sure, a big step to take. But then it felt right, this is what I am.
>
> *Anna:* It is a long process you are describing.
>
> *Ayşe:* Very long process.
>
> *Anna:* What impelled you after all?
>
> *Ayşe:* Well, I actually do not know. It was an inner drive. Since my teens I had decided that I was an atheist and socialist. But then nevertheless . . .
>
> *Anna:* What drew you to the Muslim organization? One way was to look them up.
>
> *Ayşe:* Yes, it was a friend of mine who thought that I could come with her and meet there. And then after the first time I got . . . wow, this was . . . it was the spiritual aspect. I understood that socialism was not an answer to all the questions here in life. I got a little older, had children and became a little more mature. Okay, it may solve the material problems. I looked at the Soviet Union and all those countries and it was a spiritual and moral vacuum. You can see that today, they do not have anything to stand on. So I do think I felt that something was lacking. So I went there and listened to the Qur'an and they explained. It was very competent women, one from Pakistan and one from Egypt. They were very positive and knowledgeable. I just absorbed this spirituality for the soul. I remember that I absorbed like a sponge. At home we had never talked about religion. My parents are not believers and we never visited the church. But now I just absorbed. It was really fantastic.
>
> *Anna:* What was it that happened there?

Ayşe: I do not know. I'm sure I have had this inner need all the time, but never understood it. I had maybe not channeled it to a religion, or to God. I have been a believer in one way or another. I have always believed in the good in people and in solidarity.

The defining experiences that lead to the conversion are described in various ways among the women. For some, such as Ayşe and Mariam, their travel to a Muslim country and the experience of a very different and, for them, *exotic* place constitute an important event. As mentioned earlier, Fatimah represents the "turning point" as a short crucial moment while watching a documentary about the atomic bomb and the fear this provoked. For Cecilia and Layla, on the other hand, there is no turning point as such that is expressed but a years-long process of reading and interacting with other Muslims. Ayşe's encounter with a different world gave rise to a lot of thinking. It offered her new perspectives on things and she started to question values and ideas previously taken for granted. Even if she did not understand the importance of the trip *then*, she can see today, knowing the unfolding of events, how this was from her point of view a crucial event on her way toward Islam. In retrospect, she can grasp and illustrate the continuity of a certain belief: "I have always believed in the good in people and in solidarity." This idea composes a significant continuity theme in her account, linking personal beliefs with culturally shared ideas about social justice and solidarity. Ayşe begins her narrative by emphasizing that she has always been interested in different cultures and that since her teens she has felt strongly about issues of social justice and solidarity. As I explore further in the next chapter, these ideas make up a salient personal model for Ayşe. What she experiences and presents as an "inner drive" and an "inner need" is translated and socially negotiated through a meaningful theme of social justice that is both personal and cultural, drawing on a Swedish socialist discourse.

No matter how the conversion is perceived and explained, in most of the narratives there are certain themes and ideas that can be followed from the beginning to the end, themes that frame the story and offer a coherent picture of the narrator and her life. Note Ayşe's words: "I'm sure I have had this inner need all the time, but never understood it. I have been a believer in one way or another." Through meaningful cultural themes, the women are able to form a biographical continuity that makes sense to them and, possibly, to their listeners. Similar meaning systems form life stories and present a desired wholeness and completeness that can be socially recognized.

Let us now look more closely at a religious meaning system that
negotiates personal experience into a religious story, reconciling the
convert's own unique spiritual experience with a general religious
understanding of the supremacy and omnipresence of God, explaining
retrospectively why things happened the way they did.

A Religious Meaning System

A religious system offers a more or less complete and powerful
cognitive framework. Narratives that are composed by a religious
meaning system are often made coherent by moral statements about
the way things are, how they should be, and the kind of person the
narrator is or wants to be (cf. Linde 1986:187).

As might not be expected from moral statements, these accounts
are not always stated as matter of facts but rather with self-distance
or reflexivity. As both Ayşe and Marianne mention, their experiences
and narratives could be understood as "reconstructions" or rational-
izations, but even if this is so, it does not really matter. Marianne's
critical and self-conscious narrative reflects clearly a particular
ethnographic knowledge created in the intersubjective encounter
between her and me, both trained in critical thinking.

> When I talk about myself and my relation to Islam, the view of the
> human in the creation of the universe is important. But that is actually
> a reconstruction because I didn't know that when I converted. But that
> is what I experience now. My heart lies in these things. In the Qur'an
> it says very symbolically about how the man has the ability to speak
> and have knowledge. I like that. The Qur'an says that one has got a
> certain confidence from God to be special. It is not easy to be
> human . . . you have to understand something. And that there is a cos-
> mic justice that is maintained. There must be some reason why people
> search for the meaning of life. I believe in a kind of divine contact that
> has influenced the religion. That there is something beyond every-
> thing. This is a big part of my identity.

The religious meaning system expressed here by Marianne has a
claim to completeness. Her "heart," her sense of self, lies in and
connects to a divine contact. Throughout the three conversations we
had, she negotiated meaning between me and her, between past and
present, and between personal reflections and religious and social
discourses. The short account mirrors her reflective awareness of the
continuous meaning-making of the conversion while at the same

time acknowledging that that does not make it less important or true. Islam is one of those meaning systems "purporting to explain most or all realms of experience" (Linde 1987:351). As a religious ideology it offers answers and directions regarding all aspects of life and living, or as Cecilia puts it, it embraces everything "from intimate relations between man and woman to political economy." By referring to a religious discourse the women are able to explain feelings and decisions taken on their way to Islam as well as offer an account of a fairly consistent life path.

In Fatimah's narrative her unwavering devotion to God infuses every description and interpretation of her life. While watching a documentary about the nuclear holocaust, she felt a deep, existential fear.

> What I was really lacking was any kind of spirituality in my life. And that is what this fear level is. It is kind of an emptiness of any kind of belief in anything beyond this life. And if everything you believe in is here, you are very vulnerable. You know, everything can happen at any moment. And there is no rhyme or reason to it. It happened that the business that my husband and I had at that time was land surveying. And it happened that the religious foundation needed a land-survey done of their property, so they asked us if we could come up.

Her visit to the foundation is described carefully and emotionally. At five o'clock she was woken up by a beautiful sound calling for the morning prayers. In the evening the same day she joined some people in a small room and did her first *zikr*.[4] Her experience of this is poignant.

> And we started to do this *zikr*, this remembrance of God, and it was just incredible to me. The experience is beyond any kind of word. To try to describe it was sitting in a circle and it felt like there was this spirit that was just . . . incredibly moving. That was my first introduction to Islam. Then about a year or so my husband and I got divorced. And a lot of why we got divorced, a lot that was tearing us apart was that he wanted nothing to do with any kind of spiritual practice at all. He really felt it was completely ruining our lives and everything, because suddenly other things were more important to me than making a lot of money, going on a lot of vacations, and you know. I was really searching for something.

Some time after this, she found out that the religious foundation offered workshops about different religious movements. She went back and attended some lectures about Islam, held by a man who would later play a big role in her life. During this visit she has a dream that eventually becomes of major importance. At this point she gives her dream a religious interpretation and in the moment of narration it represents a culminating experience. While telling me about the dream she is strongly moved. The emotion-laden aspect of the narrative is triggered by the correspondence between the dream and the actual experiences she had the following day. As I explore more in the next chapter, Islam is incorporated into her personal identity through a spiritual model. The religious symbolism "talks" to her; to her the message was clear and simple. She "knew" that she had to convert. From her point of view, this was meant to be.

With phrases such as "I happened to meet this friend," "it happened" that she and her husband had a business at that time, and "it happened" that the religious foundation needed that kind of service, her narrative depicts things more as events *destined* to happen than as products of her work as an active agent. It seemed to me, as a listener, that she believed that she did not have a choice. There are some parts in the narrative where she does not use personal agency, as most speakers do, but instead describes herself as following a path that is already determined and settled by a higher power. From Fatimah's point of view, she was given signs that in a miraculous way corresponded to her subjective needs and feelings. Her narrating "I" is active while her narrated "me" is passive.

Fatimah's experience of the resemblance between her inner images of the boy with a white sword in her dream and the actual encounter with the same boy the next day constitute here something similar to an "impression point" (Stromberg 1985), the moment when the religious system is given meaning on a personal level. It is internalized and as a result the religious language and ideas merge with personal experience. At this point the religious meaning system and the religious symbolism interpret not only the convert's present feelings but also her whole life story (ibid.). At certain points in Fatimah's narrative she shows how previous feelings of fear and vulnerability and lack of meaning are replaced by a coherent awareness of separate incidents in her life.

The narratives serve as a salient means to negotiate personal and emotion laden experiences into a story that can be socially shared and accepted. Through shared meaning systems, such as religious beliefs and social and political values, the women explain and communicate the reasons why they converted.

The Conversion Evokes a
Conversion Narrative

The conversion supposes not only a transformation of self-identity, as is explored in the following chapters, but also of the convert's biography. The conversion experiences, as well as the Muslim belief, must be integrated in the convert's life story. With a strong desire for self-exploration, for formulating a meaningful explanation of the conversion while at the same time obtaining a sense of coherence, the conversion narrative is forcefully motivated. Powerful conversion experiences impel the women to communicate the meaning of their conversion. In other words, the conversion requires and evokes a conversion narrative.

I have come to understand the act of becoming a Muslim as a continuing process of reconciling the change with a coherent self-presentation, achieved, among other things, through the particular conversion narrative given during the interview. This ongoing reconciliation comes about through dialogue, inner as well as outer, both through introspection and through interaction with the surrounding world. As I have shown, the converts negotiate meaning between self and other, between then and now, and between personal and cultural realms, through the conversion narratives.

The *talk* about becoming a Muslim, putting words to a major psychological event, and the negotiation of meaning after new experiences are consequently critical aspects of the conversion itself. There is a double performance of the conversion in the conversion narrative. The narration composes a creative moment in which the convert represents as well as relives experiences attached to the conversion. For example, the way I understand Fatimah's narrative is that she relives certain emotional moments while narrating. On the other hand, I have also stressed that the narrative is not the only means of self-making. In my analysis I proceed from and rely on the women's stories or talk, but I do not understand them as entirely constructing or revealing the formation of self. It is a possible, yet important, medium through which the women can make sense of life and themselves to others. As transitional phenomena they meet the desire of the convert to mediate meaning to me (and you) and the imagined audience they have in mind, to form coherence between separate incidences, and render personal matters into recognizable meaning systems.

Narratives, as well as meaning systems, operate both on a social and a psychological level. In the next part of the book I further analyze the psychological dimension of these meaning systems. How salient are these ideas to the women's self-understanding and how are they integrated with already existing ideas? By looking closer at personal models I demonstrate personal worlds of "thought-feelings," which have remained important throughout the conversion.

Part II
Personal Worlds

Chapter Four

Personal Models of Spirituality and Social Conscience

In this chapter I analyze the women's identity formation and the process of cognitive reconciliation, so prominent in the case of conversion when a new religious system is internalized and incorporated into a preexisting cognitive framework. I focus on the inner organization of a Muslim identity and the psychological appropriation of Islam through certain *personal models* and the meaning the women assign to different cultural and religious representations. This reflects a dialectical movement between the individual and the world, the tension and interaction between mental and public representations. The formation of the convert's identity occurs parallel to the gradual shift in her worldview. The salient models discussed here define not only parts of the women's perception of reality, but also their sense of self. By looking at the personal models of the converts we can gain an understanding of the personal appropriation of a religious belief and the simultaneous, and inseparable, cognitive processes of change and continuity.

When conveying the personal meaning of becoming Muslim, the women draw on various experiences, memories, and particular understandings of Islam, gender roles, family, and social politics. Some women talk about conversion and the sense of self mainly through spiritual ideas while others particularly engage ideas on gender. The women who internalized a religious worldview in their early childhood seemed more likely to describe and understand their conversion and themselves through a religious model. This chapter is divided according to these separate themes, beginning with *models of spirituality* followed by the *model of social justice and solidarity with the Third World* and continuing with *models of gender* in chapter 5. In chapter 6 the meaning of the veil is explored.

Each of the eight women[1] is discussed around a certain model; some women, however, such as Fatimah and Mariam are discussed in the chapters on both models of spirituality and models of gender and family, since both themes were highly articulated in relation to their self-understanding. Here, one cannot overstress the point that the woman I present, for example, under the model of social conscience, naturally has a spiritual sense of Islam too, but her religious identity is closely intertwined with particular ideas about social justice infused with unique personal experiences and memories. The conversion has an obvious religious dimension to it, but by highlighting social and gender models I want to show how the transformation, and continuity, is understood through other sets of ideas as well. I do not intend to simplify the women's self-presentation through the chosen division of models. That is, the converts cannot be reduced to these models. Rather, my intention is to show the particularities of and differences in the women's understandings and how a religious system is incorporated into a personal life.

In the analysis of some of the women's accounts, I explore the cognitive process of conversion by distinguishing *cognitive recognition* and *cognitive reconciliation*. These aspects of the process demonstrate the psychological appropriation of the religious system, triggering modification and strengthening of an already cognitive framework, as well as a rethinking of the self. The conversion to Islam is prompted by recognition, that is, the women are drawn to Islam since the belief appeals to already existing ideas and meets certain personal quests and desires. Within Islam, the women identify ideas that address preexisting thoughts and interests. This recognition inspires also certain actions. By attending to the idiosyncratic trajectories to Islam, the chapter highlights how specific ideas, feelings, and wishes trigger different mental versions of Islam and how experiences and ideas are continuously reconciled with previous ones. Some presentations of the women below are longer than others; this reflects the length and depth of the interviews and shows that I have chosen to present and link certain general theoretical discussions in these presentations rather than elsewhere.

Mariam—A Religious Transformation "in the Field"

In my first encounter with Mariam, I was struck by her peaceful and friendly face. I interviewed her in her home, which that day was awash with sunlight, furnished with pieces from far-away places. Since her children were at home we went to the room farthest away from the kitchen and its noises. She sat down on the bed and started to tell me the story of her life. Having been a graduate student in anthropology herself, she knew what it was all about; in that sense we shared a common world. It seemed clear to me that she had reflected a great deal upon the different alternative directions and choices in her life. Her well thought-out answers reflected a quite mindful person. In view of her profession, she has probably looked at her own life with the eyes of a critical researcher. Nothing is as given as it might seem; nothing is as simple as it might seem. During her anthropological fieldwork in northern Africa she had an experience she could not question or dismiss, a feeling of something "real."

> *Mariam:* When you go back and look on your own life, when anybody talks about their own life, it is like everybody had a point and a purpose and there was a direction. I do not really believe this. I mean, I do and I don't. I think you are guided yes, there is something that is written for you but I'm not sure it has exactly the meaning that we put to it. The way I think of becoming a Muslim, for me it was a solution . . . I imagine that is how people feel, that it is a solution to a problem that they have been living with, either of not understanding the religion that they had or combining a certain understanding that they haven't been able to combine.
>
> *Anna:* You said that when you did the prayer it felt right.
>
> *Mariam:* Yeah, for me, because it wasn't an intellectual experience. I couldn't deny it and there was no way that I could do my normal "oh, maybe it was this." No, I really felt it was very real. I'm happy for that.

During Ramadan, three weeks after Mariam had arrived at the site for her fieldwork in northern Africa in 1971, she, as a participant observer, joined the people she studied by fasting and praying herself. She then experienced something she could not reject from an analytical stance. We can explain and give different meanings to her feelings depending on personal and theoretical standpoints and preferences. The reader may have one way to explain it, while I have chosen a different model from which to discuss it. The meaning Mariam assigns to her own conversion is that it was a "solution" to particular questions and experiences from childhood.

Mariam, born in New York City, was raised as the only child in an intellectual, spiritual, upper-middle-class family. Her mother, with a Jewish background, had converted to Christianity while her father, who was twenty-five years older, was Catholic or rather "an excommunicated Catholic since he had been divorced twice." Both of her parents were personally interested in religious matters. Mariam went to some of the best schools in the country and showed an early interest

in the humanities and arts. Until high school she went to some of the
most progressive all-girls' schools where the students "were raised to
do something." In the educational program the role of pursuing a
professional career was stressed far more than the role of becoming
a mother and a wife. (I return to this discussion in next chapter.)

For Mariam, religion was an important and manifested aspect of
life. Her spiritual self-understanding is linked to particular highly
emotional memories from childhood which are significant to turn to
in order to understand the specific personal meaning Mariam assigns
to Islam.

> *Mariam:* . . . I do not think my father ever . . . my father had a real
> spiritual sense. I mean he used to tell me about the angels he had
> seen at his aunt's and about Mother Superior. It was experiences
> that were passed on from his mother to him. And my mother, I think
> she tried to be happy in it [Christianity] but in the end it didn't satisfy
> her either. I remember a nun who had been to my school taught us
> catechism. And she had taken the vow of chastity and I remember
> asking her that, "What do you do when you walk down the church
> and you are supposed to kneel down and pray—what are you
> supposed to say?" And nobody really had an answer. You can say
> anything, you just talk to God [laughs]. There was no [silence] . . . It
> was one hour on Sundays, you entered the church without any
> meaningful reason for doing that than the agony of being there. And
> I really felt that what I needed was in Islam. It wasn't a Sunday
> thing; it was all the time for everybody. Remembered all the time in
> every word and in every action, it is five prayers a day that remind
> me of the presence of God, my position in relation to him and that
> was what I wanted. If there is a God, if this is real then it should be
> important, very important.
>
> *Anna:* Were you a believing Christian when you went to the field?
>
> *Mariam:* Hmm [thinking] . . . Yeah, I think that in high school,
> mainly because my high school did a performance of Saint
> Matthew's Passion. It was a charity benefit we did in Carnegie Hall
> actually. We did the boys' choir and we spent a whole year practicing.
> I mean it is a very emotional piece. I think that I really developed a
> love for Jesus when I grew up. And always in my whole life I had a
> love for Mary. As a little kid I used to dress with a long veil, that my
> mother had, carrying the [inaudible—I think she says "baby"]
> around [laughs]. So I think I really had a connection, but not to the
> church.

Mariam's sense of spiritual connection is intimately associated to her
close relationship with her father as well as the powerful memories of
dressing herself up as Mary and performing Bach's *Saint Matthew*

Passion. As a young girl, Mariam internalized a salient *model of spiritual understanding and connection*, which spurred the appeal toward Islam, and later Sufism, and permeates the way she understands and grapples today with life as well as her memories.

After college Mariam started her graduate studies in anthropology. She found herself drawn to the area of the Sahara, and after two years she went, at the age of twenty-two, to do fieldwork in a small oasis in northern Africa. With a rather romantic and naive image of the remote area of which nobody knew anything and to which "great anthropologists could go and come back alive," Mariam saw the fieldwork as a challenge. I believe her experiences of this faraway place, and her encounter with a very different lifestyle and people practicing Islam, engendered a heightened awareness of opportunity. Her stay in the small oasis let her think beyond the ordinary identifications and categories and triggered an experience of new possibilities, a space of imagination and invention allowing for other ways of being and defining herself.

Mariam encountered and acquired Islam, not in a familiar setting but in a place where she was an outsider, a visiting Western anthropologist. Converting to Islam was also a way of "converting" to their lifestyle and an expression of her desire to understand them better, to see the world through their eyes. Adaptation to the lifestyle of the people under study is not such an unusual phenomenon within anthropology, which is reflected in the frequently used concept of "going native."[2]

Mariam describes the transformation as emotional rather than intellectual. During her three years of fieldwork she did not read any scriptures but learned the religion through the practices of everyday life.

> I became Muslim differently than most American Muslims. In some ways I feel that I became Muslim as a child. Because I took it on with learning the other new things. And it just fit and I kept it. And because the people that I was living with lived like the Prophet. They live in the same environment. Their personhood is the same. A lot of the things that I thought were customs particular for them were in fact just *Sunna*, the practices of the Prophet, that they had taken on that were natural to them. So I got a very good education about Islam without ever getting told about anything.

Mariam internalized the religion into her own personal cognitive framework, not so much through discussion, reading, or lecturing, but by *practicing* and witnessing a new cultural and social lifestyle from the beginning, just "as a child." She even found the Qur'an at that time "inaccessible." She started to see herself as a Muslim in a different and remote world where she not only had to learn how to talk, but also how to eat, how to cook, and how to pray through nonverbal communication. Plenty of what we learn and know, particularly the often taken-for-granted practical activities in everyday life, are often visual, sensual, and linked to performance. Mariam acquired Islam by participating in their everyday life, of which religion was an integral part. Her initial understanding is not really linguistic but rather based on visual imagery, embodied memories, other sensory cognition, and memories of sensations (cf. Bloch 1998:25). Mariam's appropriation of the faith is emotionally linked to strong images of faces, landscapes, and chunks of memories of smells, tastes, incidents, and certain powerful feelings linked to her fieldwork. Moreover, what she experienced of the praxis of Islam, the values and ideas of the "seminomadic" people in the middle of the desert as the participant researcher had little relevance to what she had previously learned in the seminars in graduate school, which had put emphasis on the structures and dogmas of the religion. "People didn't know Islam and for the professors it was all about inheritance, patrilineal, it was very, very strange. And when I became Muslim in the field it had nothing to do with anything of that."

Mariam had her first religious experience three months after she arrived in the field but stresses that she did not become a Muslim until after the first year when she walked for nearly a year with some of the people she studied "through the mountains proving myself to them that I could live the way they lived and get their confidence." Her intentions were to go back to the United States and finish her dissertation and then return to her new home. She expected to continue to be a part of them. "I wanted to be that anthropologist who has the same field site for over thirty years." She stayed there for three years and when leaving she felt that she was of marriageable age for them. She had learned their way of living and she estimated that in their eyes she was, in regard to her knowledge of their culture, language, and religion, about fifteen or sixteen years old. If she had received the right proposal she would have married and stayed in the small oasis. But life took another turn.

Coming back to the United States involved difficult adjustments, and she found it very hard to be a Muslim without a community. "I had no idea how to be a Muslim in America. I had only been a Muslim in an African village." With the difficulties of not having any social support validating her identity, she believes she would have left Islam if she had not met her husband-to-be at that time. They had actually met once before, three years earlier, at a time when neither of them was Muslim. Now, years later both had converted to Islam, each in their own way. Mariam's husband, with a Jewish background, had read a lot of Sufi material and was part of a smaller circle of Sufis. Mariam started to read about Sufism[3] and spirituality. "I learned what I didn't know about *Sharia*. I did more reading. Our friends that we met a lot were converts and we had a lot of talks and conversations." Mariam and her future husband did Ramadan together and shortly afterward they married. They both felt a strong need for spiritual guidance on how to live in America according to the Sufi teaching—a year later they came into contact with a Turkish man who today is the sheik (a spiritual leader) of their Sufi order. "The way things were, we really needed a sheik. We needed to see how things were done properly. It was not something you could look up in a book. You had to see a living example of what Islam really is and taste the real flavor."

At the time of the interview, Mariam and her family were part of a smaller Sufi community of around forty men and women with children. She is today a mother of five children, busy running a home school[4] for her children since she does not trust the public school system. She never finished her Ph.D. Mariam made a choice, and the field notes, still in the backpack that she carried around for a couple of years "in guilt," are somewhere in the garage as a reminder of a far-away place and an alternative course of life. For Mariam, leaving the field meant leaving graduate school, and as time passed by, the remote oasis and its people appeared farther and farther away.

The *personal model of spiritual understanding and connection* imbues her life story as a leading, powerful theme. Mariam's spiritual connection to and belief in God, acquired in her early childhood, constitute a psychological sensitivity to Islam and Sufism. Doing the prayers in the middle of the oasis evoked a feeling of having a void being filled. Through cognitive recognition, she identifies and is attracted to the all-embracing, ever-present role of the religion, reminding her of the spiritual world in which she lived as a young girl. Mariam was drawn to how Islam was integrated in everyday life. As she stressed, "Remembered all the time in every word and in every action, it is five prayers a day that remind me of the presence of God, my position in relation to him, and that was what I wanted." The salient *model of spiritual understanding and connection* triggered the conversion; the religion appealed to and addressed preexisting ideas and personal quests. Later when I asked her if she had had the idea of converting to Islam before she arrived at her field site, she

recalled a music record by Martin Lings[5] that made her aware not
only of her love for God but also of an unsatisfied emptiness in life.
She listened to this record every day for a year, long before she went
to the field, and found the recitations from the Qur'an and the calling of God's name very beautiful. Mariam thinks of this record as a
"guiding hand," "I think I fell in love with God, from that."

Mariam links the spiritual dimension and messages of Islam and
Sufism to the spiritual experiences from her childhood. By the same
token, in the process of appropriating Islam, the new belief seems to
connect to and explain previous feelings. Through her Muslim identity
her feelings about a religious connection with God and her spiritual
sense of self are reinforced. Through the conversion, the model is
confirmed and strengthened as well as modified in the process of
assimilating new experiences as a Muslim. That is, while appropriating
Islam and Sufism into her personal identity she also refashions her
preexisting religious ideas.

Through continuous meaning-making, manifested during the
interview, Mariam makes sense of her childhood experiences, her life
in the northern African oasis, and her experiences as a Muslim in the
United States. The conversion has not meant that she had to break
with her background but rather that the religious stories her father
had told her and the spiritual understanding that he passed on to her
could be integrated and attain meaning within her new belief system,
reflecting the very work of cognitive reconciliation. The spiritual meaning she gives her conversion and her Muslim identity is, as mentioned
above, that it offered a "solution"—a solution of combining her own
religious experiences with her Catholic father's "spiritual sense" and
her mother's Jewish background. "For me Islam combined the two
and I could feel comfortable with both without betraying either,
accepting both of them." Through cognitive reconciliation, Mariam
brings together and reconciles different understandings and ongoing
experiences, uniting the child dressing up as Mary with her present
Muslim identity, as well as her parents' spiritual life paths with that
of her own.

Directives and goals learned through the process of socialization
are sometimes experienced by the person as a need and a commitment
to achieve these. Mariam's feelings of spirituality were internalized
through a particular emotional relationship with her father. Her higher
aim to find and develop a spiritual understanding (like her parents')
can thus be interpreted, in D'Andrade's words, as "socialized-in
motivation" (1984:98). These kinds of understandings become "the
most general source of 'guidance,' 'orientation,' and 'direction' in the

system" (D'Andrade 1987:12). In the next chapter I show how this aim in Mariam's case existed parallel to another internalized message, the goal to have a professional career.

The interplay between the subjective world of Mariam and the religious ideas of Islam and Sufism engenders modifications in both her previous understandings and the religious discourse she has appropriated. She has internalized Islamic ideas, which incite an inner feeling of transformation as well as behavioral changes, such as doing prayers, living in accordance with *Sharia*, and wearing a veil. By studying the personal model and the psychological appropriation of Islam we can observe both reproduction and variation. The meaning-making process of becoming Muslim involves not only a reproduction of a Muslim discourse, of dominating values, rules, and representations (as we will see below in the discussion about the images of men and women and their roles and obligations) but also a generation of *personal meaning* of it, that is, a particular mental representation of the religion.

Fatimah—The Awakening of Spiritual Yearning

That night I went to bed and I had this dream that was just really incredible. I woke up in the morning and I *knew* that I had to convert to Islam. That this was definitely my way. So I went to tell him [a man who held lectures about Islam and Sufism at the spiritual center where Fatimah was staying]. Well, he said, "You are lucky. It happens that there is one other man here today, you have to have two Muslims present to be witnesses." And they were here just because his son fell of the tree and broke his arm. So he told me basically to go and make the ritual cleaning and put on clean cloth and come out after lunch and he would do it. He was very nice because he made it to a very special moment. So when I went in after lunch and into the room, there was the boy who had broken his arm and he was the boy that had been in my dream the night before. There had been two children in my dream and he was one of them. And in the dream he had been carrying a white sword. He came into the room with his arm in a sling carrying a white plastic sword. So you know I thought this is definitely what I need to be doing.

Fatimah's eyes fill with tears and her voice trembles as she tells me about her dream. We are sitting in her kitchen with a view over the small Californian town where she lives with her husband and two children. Like Mariam, Fatimah was drawn to Sufism and she and her family are part of a Sufi community. Looking back on her major decisions and turns in life, she seems today to be at a point in her life where she feels fulfillment and peacefulness. Fatimah had the dream after a couple of years of spiritual yearning while visiting a religious foundation.

The resemblance between the "real" world and the dream world has strong emotional impact on her; she understands it as guidance from God. As I discussed in chapter 3, the analogy between images and creations in a perceived inner world and the actual encounter with the same boy in real life is given a religious interpretation. Fatimah's understanding of the dream has immediate consequences in her life—it triggers her decision to convert. The importance given to the dream and the fusion between dream and reality reflect Fatimah's personal model of spirituality, an inner personal world of "thought-feelings" on spiritual connection and yearning linked to her emotion-laden memories as well as her self-understanding. Corbin (1966) describes how the Islamic Sufi tradition elaborates an "imaginal world," a form of reality where sensibility and spirituality meet. Contrary to Western thoughts of oppositional dualism of reality and imagination, between physical and spiritual realms, the Sufis describe and explore the imaginal world as a real world of autonomous images.

The construction of meaning of the dream begins in the transformation of the internal image experienced, the visual imagery, into its translation in "language-thought processes," the cognitive categories of the dreamer (Edgar 1994:100). Fatimah interprets her mental images of the dream not only through preexisting religious ideas but also through her strong yearning for transcendental meaning. Her dream reflects something about the current organization of self, the emotional situation she was in, and what was of cognitive and emotional salience to her at the time (Hollan 2001).

Fatimah's understanding of her dream resembles somewhat the thoughts of the psychiatrist Carl Jung who understood dream imagery as symbolic representations of the dreamer's future. The dreams are interpreted as having a prospective function, offering an opportunity to determine future goals of the self. By reflecting on our dream imagery we can, according to Jung, gain crucial insights that could lead to actual changes in life (Edgar 1994).

Fatimah integrates her dream experience in her personal life, and in relation to the emotional situation she found herself in, it gains strong emotional and motivational force. For about three years she had been on a "spiritual path." (Just like Mariam she uses the concept of "spiritual" frequently, a common and fundamental notion within the discourse of Sufism.) In this situation Fatimah is highly sensitive to any possible message that could be given religious meaning. In her interpretation of the dream she works out a resolution of a current unsettling situation and personal conflicts; her understanding of the powerful image of the boy with a sword and her re-encounter with the same boy the next day has a problem-solving capacity. Spiritual yearning and the strongly felt urge to find a resolution to her anxiety reflect Fatimah's appropriation of Islam.

Fatimah had the dream during her third visit to a spiritual center. Her initial contact with the center, some years earlier, was after watching a film about nuclear holocaust, a film that evoked a powerful experience of emptiness and fear (see chapter 3). "This fear was at such a deep level that I didn't know what to do with it." Soon after she met an old friend who lived at the foundation.

> He gave me some books to read and I think there was one about Taoism and one about Zen Buddhism and that kind of got me started. What I was really lacking was any kind of spirituality in my life. And that is what this fear level is. It is kind of an emptiness of any kind of belief in anything beyond this life. And if everything that you believe in is here, you are vulnerable. You know, everything can happen at any moment. And there is no rhyme and reason to it.[6]

Since she and her then husband ran a business together offering services that the religious foundation needed they made a deal. They would do the service for free in return for coming to some of its workshops. This offer suited Fatimah's needs and interests well. This visit was the beginning of her "spiritual journey" and her first encounter with Islam.

> The place we were staying in was a series of rooms that had been built like little adobe-mud rooms that were almost built in the style of a Moroccan street or something. It was in this kind of semicircle and it was a really tiny little space. And then outside, it was in the mountains, and there were no people living close to the foundation. So we were in this environment and I woke up about five o'clock the first morning to the sound of the *adhan* being given, rolling down the mountains, just this beautiful *adhan*. I mean it was just as my heart went pooaah, what was that? It was just such a beautiful sound. Someone had gone up and was giving the call for the morning prayer for people there. Maybe only one of them was

Muslim but the rest would come each morning to pray. So it turned
up that it was my friend who was giving the prayer. So I asked him
about it, you know. And he told me that, "actually we are going to
have a *zikr*." Do you know what a *zikr* is?

Anna: Not exactly, some kind of ceremony?

Fatimah: Zikr translates it to mean "remembrance of God." You
come together and you recite over and over the names of God, and
"there is no other God but God." And it is a kind of chanting, or
singing combination. It can be really powerful. And I went and there
was about nine or ten people in this little tiny room. And we started
to do this *zikr*, this remembrance of God, and it was just incredible
to me. The experience is beyond any kind of word. To try to
describe it was sitting in a circle and it felt like there was this spirit
that was circling the room and it kept going up and up and taking
us with it. And it was just . . . incredibly moving. That was my first
introduction to Islam.

Some months after this visit she got a tape with prayers and a
handout of the positions for the prayers from the same friend
who had given her some books earlier. Then, for a period of a year
and a half, Fatimah did her "own little ritual" alone. Every morning,
before the rest of the family woke up she meditated and did the
prayers. "And then I would get out running and come back and face
the day [laughs]. And that was all I had. At that point I wouldn't say
that I even had converted to Islam. It was just a practice." Fatimah
did not know anything about Islam, she did not know what the
words that she said when she prayed meant, and she did not know
the doctrine or any of its history. She continued doing it in solitude
since it meant something emotional to her. She "loved" the practice
but we cannot say whether that was a Sufi or an Islamic practice to
her at that point. She was saying words that she did not know the
meaning of, to which she had no reference. So what meaning did it
have to her?

I mean I was still very Catholic in terms of my connection through
Jesus, peace be upon him, you know, through the whole Christian
teaching. And I didn't know if Islam was similar or different or
anything. I didn't know anything of the history. And I just knew that
I loved the practice that I had been introduced to so far.

Fatimah understands it through already learned Catholic ideas about
a presence of God, "something that was very close to me." This was
during a very difficult period in her life. Fatimah's interest in the

spiritual practice had also started to tear her and her husband apart and they eventually got divorced.

When she was asked to come back to the center again to participate in a weeklong workshop, she gladly accepted the offer. This time she met a Sufi teacher who would become very important to her. There were two messages that had great impact on her, these addressed ideas that she had been struggling with earlier within the teaching of Catholicism. First of all, that there was no guilt in Islam, which was "a very releasing sort of thought." Secondly, "in the Sufi teachings there is said to be as many ways to God as there are breaths of creatures. And that any one of these paths ultimately leads to the source. Some are straighter than others. And this again was very different from anything in Catholicism, where you had to be a Catholic or you couldn't get to heaven." On the other hand, she also found continuity between her religious background and Islam. During her stays at the center she "soaked in as much as I could about anything to do with Islam."

> So all of a sudden I didn't have to give up what I had been taught as a child: about Adam and Eve, about all the stories of the prophets, the story of the saints, the story of Jesus and his mother. Instead it was like as it became richer, details got filled in that had never been there before, that pulled it all together. And suddenly, it was like having somebody open the door to a wider world instead of closing off saying: if you love this kind of praying you have to give up what was dear to you before. It was like you could embrace them both.

Once again, conversion is defined and understood not only through the changes following from becoming Muslim, but also through continuity, reconfirmation of preexisting values, and an all-embracing feeling of connecting earlier understandings with new ones. Like Mariam, she did not feel as if she had to give up strongly held values from her childhood. Through cognitive reconciliation, Islamic and Sufi ideas are internalized and assigned meaning through certain Christian ideas, stories about the saints, and her love for Mary, reconciling "old" and "new," reinforcing a new sense of a religious self, a Muslim identity. Furthermore, Fatimah's experience of a powerful connection to Mary reflects a deeper personal identification with motherhood, which is a crucial aspect of her self-understanding. The new name given to her after the conversion symbolized and reinforced this experienced attachment, linking past with present.

> My name before I was Muslim was Fanny, actually it was Frances, which I liked, but I grow up hating Fanny [laughs]. So I was eager to

change my name [laughs]. I would like to take Miriam cause I felt very connected to Mary and to the image of the mother so when I went to convert I asked the man who was teaching the class if he would give me a name. He said that when he had converted in Sudan they had the tradition to take the first letter in the name and give them a Muslim name using that first letter. So he gave me the name Fatimah. And when I asked him what it meant and he said that Fatimah refers to an aspect of Mary.[7] So suddenly there was another affirmation that I definitely was on the right path. So I slowly, slowly started . . . I didn't tell anyone.

Similar to her interpretation of the dream, the meaning of the chosen name was interpreted in spiritual terms, that she was doing the right thing. The correspondence between personal feelings about motherhood and sentiments attached to the image of Mary Superior, on one hand, and the meaning of the Arabic name "Fatimah," on the other, is far from being understood as a coincidence but rather as a spiritual sign.

The period after the official conversion, when she left the spiritual world at the center for her own, was probably the most difficult time period in the process of conversion. She was a Muslim officially but with hardly any insight into the religion. As is so poignant in the case of Fatimah, converting requires a set of social relations and a supporting context. She had neither. As we see in the next chapter the importance of a Muslim life partner in order to implement the new belief and change in life was stressed by many of the converts. Fatimah had just gone through a divorce. She was living alone, working and sharing her time with the children with her ex-husband. Something fundamental had happened to her but she had difficulties putting it into practice. Fatimah reflected upon her spiritual problems but she had not acquired any Islamic words for it. She was scared and felt lost in a world where she did not know anybody to relate to on a spiritual level.

So in the end of that month I went back and my life completely fell apart. And I was living by my own and working. I had the two kids and it was really hard and just . . . in painful ways that I do not really want to go into. And I just stopped doing everything; I stopped doing any practice. I now had taken *shahada* and I totally dropped it. Decided that . . . it was almost like a reaction like, no prayers . . . I didn't know any other Muslim that was the other thing. I had no association with anyone. All I had was this little tiny kernel of information [laughs]. Not even knowledge. And I had ordered a copy of the poetry of Rumi and I kept telling myself that when that came I was going to start praying again. And it didn't come for months and months [laughs] and I kept calling them and they said: "Yes, it is coming, it is coming." And I couldn't get a copy in the U.S. Somehow I couldn't pull myself together and pray.

Anna: And there was nothing else in life that confirmed that you were Muslim?

Fatimah: No, not at all. Quite the opposite. It was very much the opposite. I was working every day, wearing short skirts and short shorts and totally, completely into American. I mean I just didn't have any associations with any other Muslims and I was afraid too. I was afraid because I had been warned that they were very strict Muslims and that it would be very hard to associate with them. I was afraid to meet anyone.

Anna: It was easy at the foundation but when you left it—

Fatimah: —when I left it, it was almost more like I was afraid of it. It was something very profound that had happened to me and I didn't know if I was ready to be that deeply involved in it. So I just stopped everything for about five months. And meanwhile my life was spiraling down hell quickly on an emotional level. You know, things were going very poorly. I knew that a lot of it was that I really was a family person and really wanted to be with somebody and it was very difficult to find anybody that I could even relate to, you know, on that level. So one day I just decided that I have to start to pray, I have to start doing the prayers every day as many times as I can. If I can't do it five times I do it twice, I do it three times. So I started then and the very next day when I came home from work the book was there on my front porch [laughs]. So that was really kind of . . . what comes first the chicken or the egg? You have to decide that this is what I want and that was how I felt anyway. That once I made the decision—

Anna: —that the decision should come from you?

Fatimah: Right. And then I was faced with . . . still feeling that my life was hitting rock bottom. I would wake up in the mornings feeling like somebody had impaled a spear through me into the floor and I couldn't get up from the bed. It was just a deep depression. The sky was gray, however blue it was. And if you ever have been to New Mexico the sky is so blue [laughs]. It was just a really, really hard, horrible feeling. One night I just reached the bottom place that I could imagine I could go and I did the prayer, the evening prayer. And afterwards I just sat down in tears and prayed: "Dear God I really need a teacher. I need not a book; I need a living, breathing master of Islam on earth. I need a Sufi master. Please. And I do not know how to get one, I do not know where to go, I do not know who to call and I'm really stupid so could you put it in black and white for me. I'm going to open this book and could you tell me on the pages of this book, which was Rumi's book, who do I call to find this?" And I opened it up to the story of this man, Bilal, and he was the only other Muslim I knew in the whole world [laughs]. That was this man whom I had converted through, who lived out here in California.

Fatimah went through a serious depression, feeling her life "hitting rock bottom." She had gone through a profound religious experience but she found it difficult to integrate this spirituality into her everyday life. The sensed gap between inner and outer provoked a deep anxiety, a "feeling like somebody had impaled a spear through me into the floor." She got into contact with the man through whom she had converted and he recommended her to get married as a solution to her depression and difficulties of living alone as a Muslim. This man knew an American who also had converted to Islam and who was also divorced. Fatimah agreed to meet him and after a couple of months they married. She describes this encounter as magical. They connected, she gained emotional and social support in further exploring Islam, and she experienced a new world opening up for her through her life companion, who also could confirm and recognize her Muslim identity.

Fatimah's identity as Muslim is organized around a salient *model of spiritual yearning*. The personal model, based on understandings about human beings' need and quest for spiritual meaning, explanations for questions about life and death, and connection with God, embodies a strong emotional commitment. Even as a child Fatimah had a religious feeling "of the presence of God." She "loved" the stories of the saints, parts of the practices she grew up with, and some of the masses that made her feel "uplifted." During her years in college, however, Fatimah "gave up religion." The church did not give satisfying answers, they had "modernized the mass" and there was no longer any "feeling of sacredness around the whole experience." For years she was preoccupied with family and work. The film she watched about the nuclear bomb some years later evoked a sudden and overwhelming feeling of vulnerability due to the lack of spirituality in her life. Edward Sapir argues that "we can expect no more of any religion than that it awaken and overcome the feeling of danger, of individual helplessness" (1949 [1928]:347). For a religious belief to awaken feelings in the first place implies, however, that the religious system has the force and affinity to "speak to" the individual or, in other words, that the individual has internalized ideas that can be activated and animated.

In Fatimah's case, the decision to convert to Islam is triggered by the *model of spiritual yearning*. The experiences of and encounter with the Sufi teachings at the spiritual center and her own ritual that she was practicing alone every morning appealed to her personal spiritual quests, prompting cognitive recognition. This was what she was looking for. Her experience of *adhan*, the sound of the call for the morning prayer, her participation of the *zikr* and her dream, all understood through a particular personal inner world, gained strong motivational force, further inciting practices such as the official conversion and praying.

Fatimah's understanding of her transformation from feeling lost and depressed to a believing Muslim is organized around religious ideas with which she can also connect her early feeling of the presence of God with whom she is today. It is easier to tell one's story when knowing the "result" of the long whirling road to Islam riddled with emotional breakdowns and depressions. Today, fifteen years later, she has not only learned about Islam and Sufism but she also belongs to a Sufi community. To return to Sapir, Fatimah feels that she has overcome feelings of danger and helplessness through her identity as Muslim. Islam has gained cognitive and emotional salience through a *personal model of spiritual yearning*.

Lisa—"Help Me, Help Me, Help Me!"

Lisa is the youngest of the women. She was in her early thirties at the time of the interviews, and she had not been a Muslim for more than three years the first time I met her. Despite our new, and temporary, acquaintance, I sometimes felt that the conversation was as comfortable as between two friends. I experienced the interviews, particularly the second one that took place in her home, as relaxed. Before and after the recorded dialogue, and even during it, we talked about many other things and I believe that with her I talked about myself more than I did in many of the other interviews. We were sitting in her and her husband's small sunny apartment, each in an older comfortable armchair eating Turkish delight.

The first time I met Lisa was at the Muslim organization that had mediated the contact between a couple of interviewees and myself. Apart from Fatimah, Lisa is the only one of the women who mentions a conscious, intensive religious searching caused by a personal crisis as the initial step toward Islam. She grew up in an atheist family where religion never was an issue of concern. Like Fatimah, her experience of becoming Muslim was filled with feelings of loneliness and uncertainty. She found little support and understanding from family or friends. However, by the time of the interview, Lisa had come to experience stability and peace. Since the period of doubt and depression she had found satisfying and meaningful messages within Islam, translatable to her own experiences and wishes, and she had met a life companion.

Like many other young women in their twenties, Lisa had boyfriends, went to discos, and partied. However, at the age of twenty-seven things changed.

> Then . . . it is now almost four years ago since I went through a life crisis. I changed lifestyle . . . I ended a relationship which was very bad and in relation to this I began to want more contact with God. And the only thing I was occupying myself with was "help me, help me, help me." It was like when you feel really, really depressed and do not know what to do with yourself. I have rather vague, woolly memories of that time.

Lisa visited the church sometimes, to light some candles. She started to pray, even if she did not really know how to do it. Some months later she began working at a school with students from different parts of the world. It was there she got to know some Muslim friends and from them she borrowed some books about Islam. Lisa found it interesting but was for some time rather reluctant and skeptical. She was torn between a growing personal interest and the knowledge of the general preconceived, negative notions about the religion. Appropriating the Muslim faith involved rejecting and reworking previously learned ideas about the religion and Muslims.

> *Lisa:* I was interested so I continued reading more and more and could not really let it go even if I wanted to. Moreover I am not that stupid that I could not figure out what consequences it would have at work and within my circle of friends and in the society [laughing]. Not only myself started to wonder if I was out of my mind but also other people felt "uh, why does she have to deal with that sort of thing?" All of my old friends dissuaded me from Islam. So I didn't really get any kind of support.
> *Anna:* No . . . did this last for a long time?
> *Lisa:* No, it wasn't for that long . . . it was a couple of months more. Then I got it into my head that I was going to go to Turkey alone, and so I did. I was already interested in Islam and I knew quite some—
> *Anna:* Let me just ask you, the time from that you started to read until you left for Turkey—

Lisa: It was about six months. So I went and there I came into contact with different families and became kind of admitted.

Anna: Where you away for a long time?

Lisa: No, I was there for only a week but it changed a lot. I got accepted there as a Muslim.

Anna: You became a Muslim there?

Lisa: Yes . . . it was still kind of a gray zone, I don't really know what I was. Some can maybe say a date that then and then I decided but that was not how it was for me. This was a slow thing that flowed out over the period of a year or so before I was really sure. During my visit in Turkey I was still uncertain but they accepted me and took care of me.

Here, Lisa touches upon the gradual process of converting, of understanding and integrating a new religious belief. It was a "slow thing," going through periods when she did not really know what she was. During her visit to Turkey, Lisa started to wear the veil. This was also convenient, as she added jokingly, since as a blond woman she got far less attention from men. In Turkey she started to explore a new world, externally as well as internally. There she found social support from the families she met; there she was recognized as someone who she was not yet ready to be in Sweden. She found it more or less impossible to go out in public in Sweden with a veil covering her hair. It made her feel like a UFO. Lisa went back to Turkey for a two-week-long stay, and it was visits such as this one that were "strengthening" during an otherwise lonely time period. Looking back during the interview she had a hard time pointing out particular things that have been of importance in her decision to convert. "Many have asked me why, but I have difficulties answering them," she says. "I don't really know myself."

Becoming a Muslim in loneliness was difficult but she was driven by a deep, inner need of spiritual contact and what she refers to as her awareness of God. Lisa had strong spiritual experiences and she remembers praying to God, "help me, help me, help me." And God did answer her prayers. The immediate feeling of God's presence, which Lisa illustrates with the Muslim saying that "God is closer to you than your own jugular," left her with an undeniable, persistent experience of an "inner reality," an experience maintained by daily, meditative prayers.

During her visits to Turkey she tried out an identity and practice that appealed to her personal quest of spiritual meaning; in the Muslim pillars of faith she found a much longed-for stability. Trying out a Muslim identity in a different setting is an interesting trait among the converts, whether it is Turkey, a small oasis in northern Africa, one's husband's African home country, or a spiritual center in New Mexico. It reflects the experience of a unique opportunity to explore new self-definitions. In a context where she is not restricted by certain expectations from people close to her who already have a fixed idea about who she is, the convert senses a longed-for and immediate opening to try out a Muslim identity. Here, self-reflection and self-exploration are encouraged by supportive others reflecting a quality of the conversion—a dialogue between inner construction of self and social recognition of the self. During the first years, Lisa had very little contact with any Muslim or Muslim organization. Realizing her female Muslim identity seems to have become easier for Lisa after marrying a Muslim man. The social support from having a Muslim partner is essential, particularly in the beginning. The other can approve and confirm the transformation, the new outlook on life, and the religious self that parents and friends might not recognize and accept. Together, Lisa and her husband negotiate their meaning of Islam; the religion forms an essential part of their relationship.

Lisa has appropriated the Muslim faith, which permeates her worldview as well as explains her previous experience of emptiness and lack of spiritual meaning. It efficiently makes sense of her memories and past. Even if she first told me that neither she nor her parents were religious during her youth, she claimed later, in the second interview, that God has more or less always been somewhere in her thoughts, but she has been "forgetful." The faith was internalized at a time of deep emotional stress, and in the process of making it into her own it addressed her particular personal needs and quests. Lisa's Muslim identity is organized and negotiated around a prevailing *spiritual model of inner reality*. She learned about Muslim teachings during a personal crisis in life and the religion was acquired through a strong yearning for change and inner peace. For Lisa, Islam is a sense of connection with an inner reality within herself.

Layla—The Journey of Life Destined by God

Layla, who is in her mid-fifties, is one of the oldest of the Swedish interviewees. In the previous chapter on the conversion narrative, I discuss how Layla organizes coherence in her narrative by connecting her early belief in God to her conversion to Islam and where she is at today. By referring to a preexisting idea about the existence of God she incorporates the conversion, a change in worldview and identity, into a self-coherent biography. Through a *personal model of religious guidance* she defines and organizes a Muslim identity. Most importantly, she has "always believed in God."

In her twenties Layla went to India together with two friends of hers. The visit had a big impact on her. "It opened my eyes since I wasn't used to seeing other cultures and other kinds of people. So it widened my perspective." When she came back she started to see a Muslim man who she had learned to know while working in a bookstore and a year later they married. Before the marriage she decided to convert to his religion, even if, as she stressed, her husband did not insist on this. At that time Layla did not know much about the teaching of Islam; it was the beginning of a long process of learning about the belief through reading and making Muslim friends. Her husband was not really practicing Islam, and during the first years of their marriage her conversion was not given any religious meaning. It was not until fourteen years later that she started to wear the veil. During this long period of time she read more and more and later she began to discuss Islamic texts with a group of other Muslim women.

Layla's discussion about *hijab*, the veil, reflects her ideas about the relation of self to a religious belief. For Layla the veil is a marker of her relationship to God, a measurement of her devotion. As she expressed it, she did not fear or believe in God enough to put on the veil until much later. Similar to Lisa, at the "first stage" of the conversion process Layla thought more of how people around her would think and react. She describes it as a slow process in which nothing outside forced her. The idea that the decision has to come from "inside" is essential to the converts. There was, as she puts it, an "inner feeling" that is hard to explain and to put in words, a feeling that she was doing the right thing.

Like the other converts Layla took on a Muslim name when she officially converted to Islam. Similar to Ayşe's reflection on her names in chapter 3, Layla unites the meaning of her given and taken names. She was given the names of Eva Lena Susanna and she later took the double name, Susanna Layla, even if she most often goes by Layla.

> *Layla:* I chose both of them because I felt that they were my identity. Susanna Layla may sound a little strange but I could not just have Susanna or Layla as my only first name. Susanna Layla—it feels like that is me.

Anna: Both the Swedish and the Muslim.

Layla: Yes.

Anna: Well, your name has to do with identity . . . it may have been hard not to have a Swedish name?

Layla: Yes, I don't think that would work. Then you kind of take away parts of your roots. They are important to keep. They have formed you and made you into the one you are.

During a long period of time Layla has slowly reorganized her self-understanding as Susanna Layla. "Susanna Layla," Layla says trying out the names, "it feels like that is me." While talking about their old and new names, the women also express the simultaneousness of continuity and change, the experience of being both the same and different. They have changed but not to something totally different. Most importantly, the convert's sense of transformation is experienced in a preexisting cognitive framework, in a particular personal world. Reading and learning about Islam confirmed previously held moral values as well as an existing self-identity, an identity she refers to as her "roots." She gradually discovered that Islam embraced values she always had believed in and approved of. In regard to drinking, partying, caring about appearance, and having many boyfriends—these are ways of living that have never seemed appealing or desirable to her. Through cognitive reconciliation, earlier religious ideas are integrated with Islamic ones; and her self-understanding as Susanna Layla unites different aspects of self and moments in life, connecting past with the present and an anticipated future.

Faith in God and the idea that everything that happens is meant to be compose a life- and self-defining model for Layla to which she can connect her "previous" life as Susanna. The *personal model of religious guidance* is invested with idiosyncratic emotional meaning and defines ongoing experiences. The meaning she assigns to Islam, intimately linked to her sense of religious self, implies being a part of something bigger, being led by a divine power, and being a better person.

I believe in fate and when I look back on my life it is like everything falls into place. That I can see that this or that event made me go this way . . . without being active myself. Of course I have . . . I don't know. It feels like it is with God's help that I got where I am at now in life . . . You try to think more positively, to live here and now. And of course that has also to do with experience that you gain. That you try . . . it is easy to be irritated or angry but that you have to be aware of trying to mitigate it. The purpose is to become a better person. I do not know. That you should not talk bad about others or slander others or be malicious. All of these human characteristics that you have within you, that you should try to improve. And this is something I try more consciously now since I believe in all this, your free will, fate, the angels and everything.

Now and then in the interviews, Layla invokes the voice of strict Muslim interpretations and standpoints that could be understood as controversial, not only by readers in general but perhaps also by some Muslims. Even if this could be the standpoint of other converts, they did not vent them as openly as Layla. In *Sharia*, the Islamic law, you can read about the approval of the death penalty. In Sweden, I would dare to say, the general public opinion on this kind of definite punishment is very negative and therefore it could be seen as a difficult issue for a Swedish woman to relate to and publicly elaborate upon. Layla is reflecting loudly on it, highly aware of the polemical nature of the problem.

Anna: Is there anything in Islam that you find hard to relate to, to accept?

Layla: Are you thinking of the death penalty?

Anna: I'm not thinking of anything particular. I just wonder if there is something that has been difficult to agree with.

Layla: There is the death penalty in Islam. And then it also depends on which countries and those in power we talk about.

Anna: Is it mentioned in the Qur'an?

Layla: In *Sharia*.

Anna: In the law.

Layla: There are death penalties for different things. In Pakistan, for example, they have death penalties for use of narcotics but if you have money you can buy yourself free [says something inaudible].

Anna: Is the death penalty such a thing?

Layla: [Sighing] I don't know. They have it in America. I don't know. [Layla talks for a while about the, in her view, far too mild punishments in Sweden]. But then you also know from the United States that it doesn't deter. But on the other hand maybe there are people who don't have the right to live on earth, I don't know. I think of those who do terrible things. Those who kill children and are sitting in jail. People who can't be helped . . . I have to accept the death penalty, because it exists in Islam. But if you look at the Muslim countries you can see that it is not applied in the right way, from an Islamic point of view.

Layla is uncertain. It is obviously not a simple question and it is not the first time she thinks about it. She is trying to find reasonable explanations and ways of looking at it, searching for her own ways of reasoning to correspond to the rule prescribed in the *Sharia*. It exists there, in black and white, and in her mind she cannot ignore or question it, not if she is to call herself a truthful Muslim. She ends her reflection with: "I have to accept the death penalty, because it exists in Islam." Even if she does not say it explicitly, I assume this was not necessarily a view of hers before the conversion. Converting to and embracing a new worldview—does it have to involve a total recognition and identification with it? The women find their own different ways of relating to and interpreting issues, at the same time I do believe it is hard to just "buy half of the package" and still be recognized as a genuine and serious Muslim, both by oneself and by others, Muslims as well as non-Muslims. Layla has acquired and accepted the idea of the death penalty as an authoritative representation, difficult to question, since it is "God's words." The way Layla understands Islam and herself as a Muslim is that she has to follow God's guidance and that she "has to accept" the issue.

Referring to a Muslim ideal of living in balance between the spiritual and the material, she expresses a future aim, her wishes to live more harmoniously with nature and to develop her spiritual mind. For financial reasons and out of a sense of responsibility to her children, though, she perceives the difficulties of transforming the ideal into practice. But if she was "given" a year, as she put it, she would go to a Muslim country and explore the spiritual aspects of existence in her life.

Layla often refers to common religious explanations, that it was an "inner drive," an "inner feeling," and a guidance from God that made her become Muslim. The religious model offers explanations as to why she converted, but it is also a personal, meaningful understanding, strengthened by particular personal emotions, of how things *are* and *should be*, an understanding that directs her everyday actions and decisions. The religious model composes ideas and emotions that are perceived as "natural" and "right" (cf. Quinn 1992), and hence specific religious norms and ideas are reproduced.

Ayşe—Social Justice and Solidarity with the Third World

Political and social ideas were always talked about more or less explicitly during the interviews. As a part of the conversion, or maybe sometimes as a consequence of it, the women criticize the

Swedish and the American society and Western values and attitudes in general. Contrary to what is often assumed, religious conversion does not necessarily have to be motivated primarily by religious ideas but by political and social understandings as well.

Becoming Muslim allows the women to explore new worlds and selves, an exploration that also involves a reevaluation of one's previous lifestyle, and in this reassessment the West is sometimes opposed to the East, Christianity to Islam, secularism to spirituality, modern to traditional, and individualism to family-oriented values. These ideas are partly built upon an older culturally constructed dichotomy between the West and the East, reflecting a form of Occidentialism. Social and political criticism of Western civilization and ideas of individualism, egoism, secularism, and superficiality have repeatedly been expressed from different perspectives by intellectuals, politicians, and other converts (e.g., intellectual converts to Islam at the beginning of the twentieth century (Gerholm 1988)). The women expressed similar general views, often with the purpose of giving a larger social explanation for their conversion, but the ideological comparison is also made in a particular cognitive framework. Ayşe in particular explored her Muslim identity and conversion through social and political ideas.

Ayşe was the only one of the women who did not wear a veil when I came to her home for the first interview. She wore black makeup around her eyes and a black, velvet-like, fairly tight dress. She talked passionately about Islam from two main premises: solidarity and social justice and the strong community of Muslim women.

Ayşe understands her conversion and formulates her self-understanding in terms of explicit political and social ideas shaped by a socialist discourse prevalent in Swedish politics. She negotiates a Muslim identity and has acquired Islam through a *personal model of social justice and solidarity with the Third World*. As a vital and well-articulated model, linked to specific emotional experience and memories, it triggered the very decision to become Muslim and it constitutes an all-embracing cognitive framework through which she interprets ongoing experience and perceives a coherent sense of self through time. She sees herself as a Swedish Muslim activist, engaged in social and women's issues.

At a rather young age, Ayşe internalized and embraced socialist values, encouraged by the political and social movements of the 1960s and the 1970s. She grew up in a secular home in which religion was never discussed, and not unusually for her generation at that particular time in Sweden she perceived herself as an atheist and a convinced socialist. In her teens she also decided to leave the Swedish church. Ayşe recalls a forceful public and dominating message that religion was something bad and religious people something you should despise. Her teacher in religion taught them that Islam was a violent religion that oppressed women.

When Ayşe turned sixteen she met the man she later would marry. They lived a life like most young people; they lived together before marriage, drank wine, and went to the disco. Her husband-to-be, who was born in an African country and in a Muslim family, was also a socialist and atheist. Ayşe felt a strong engagement for the Third World, and a fascination for other far-away places and cultures. She traces this interest to her early childhood. Ayşe's belief in social justice and solidarity with the Third World are infused with particular childhood memories and experiences in adolescence.

> When I was seven years old I read a lot and drew Indians and Indian tents and Africans and huts. Yes, you know all these stereotyped images. I have been interested in foreign cultures right from the beginning. And especially Africa, which fascinated me when I was very small. Then I actually met my present husband from Africa. We lived together before we got married. And there were no discussions about Islam and I was in the socialist period of my life then. You know it was during the seventies and it was the U.S. out of Vietnam.

In 1975, at the age of twenty, Ayşe made her first trip to her husband's home country in Africa, which opened her eyes to a new and different world. This was also her first encounter with Islam and Muslim people and practices.

> I was very impressed by the people there and how they lived. They lived a poor and tough life but it was just wonderful, warm, and friendly people with strong family solidarity. I saw these big families and the community. Wow, it went straight to my heart. And I arrived there as an ordinary Swedish teenager with long hair and a miniskirt who knew nothing about Islam. Islam, what is that? I was totally against Islam, I was an atheist. That I had decided and I had left the Swedish church. But I really felt like I had come home. Africa, the music and the dance. I soaked in everything. It was a revolutionary trip for me.

Ayşe integrated the experiences from the trip into her worldview colored by her political engagement in the "Third World" and left-wing radicalism. It was a "revolutionary" and "magical" trip in the sense that she got her own personal experiences of a so-called developing country and of practices to which she could relate her ideals. The social system of taking care of your family, relatives, friends, and neighbors and "the warmth and solidarity between people" was captivating. She "felt at home." Thus, it was not really the religious message that she found enchanting at first, but the exotic "otherness" in the African culture. It started with a young woman's attraction to African clothing, music, and traditions such as painting henna on her hands. Furthermore, the strong, confident women and what she perceived as secure female identities amazed her.

The visit, however, triggered thoughts about the religion and Muslim practices, but it would take almost ten more years until Ayşe came into contact, through a friend, with a Muslim women's organization in Sweden. A couple of years later, in 1986, she made her second trip to Africa and this time she symbolized the change she had gone through since her last visit by the length of her dress. This was the beginning of five more years of reading, meeting other Swedish women who also were considering converting, and participating in seminars and lectures held by different women's organization. Once again, she was caught by surprise and intrigued by the confidence of the Muslim women.

> These were practicing Muslim women, who were very knowledgeable, and that didn't at all fit into the clichés of the Muslim woman as illiterate with a kerchief and ten kids around her. Instead these were educated women who had answers to all kinds of questions. It was very fascinating and stimulating.

During the five years of reading, she questioned a lot and looked into what Islam stood for on different issues. There were many obstacles in her way, preconceived notions not only about religion and religious people in general but about Islam in particular. Also, she found it hard to put the word "God" in her mouth because of the "history of Christianity and the slave trade." It was easier to say "Allah."

Ideologically, converting to Islam meant being one of "the others," not an ordinary Swedish woman, but someone standing for and identifying with a religious system that, as she perceives it, represents an opposition to and rejection of Western values (read: materialism, individualism, and lack of spirituality, warmth, and a sense of community). As a Swedish and more generally as a Western woman she saw herself being on "the rich side in history." As a Muslim she could, however, identify with, have loyalties to, and be a part of the other side.

> The Swedish church was so old and obsolete and there are these missionaries who want to force their own civilization on others. I was politically a socialist, supporting the third world, against the war in Vietnam and against the Western imperialism. And these ideas have mattered a lot for why I was pulled toward Islam, which is so much about being against Western imperialism. Islam stands for everything that is not Western . . . there is a kind of state of opposition. To me Islam represents the Third World and another perspective and world-view than the Western materialistic one. So, I believe that played a big role in my development. That I felt at home in Islam.

Within Islam Ayşe could further explore and elaborate her political sense of self and discover new ways of understanding her place in the world. The *model of social justice and solidarity with the Third World* organizes Ayşe's identity as a Muslim. She stresses at one point during the first interview a significant experience: "I'm still the same." She understands and interprets the transformation (from being a liberated young woman, a socialist and atheist with a short skirt, believing in free sex, to a veiled Muslim woman) through these ideas of social equality and solidarity linked to emotional memories such as the trip to Africa. The personal model is powerful since it has been acquired during her early teens and later intensified and recon-firmed, as well as modified, by her commitment to Islam.

Ayşe expressed her difficulties with stricter interpretations of Islam and certain religious restrictions. She has, in her view, adapted a rather "liberal" version of Islam. The Islamic practices she encoun-tered in her husband's home country in Africa fit nicely with her feelings about dancing and music and a relaxed attitude toward relations between men and women. In that regard she "converted" as much to the African culture that she got to know from her visits and through her husband.

> It is a more tolerant interpretation, not as rigid. It suits me fine anyway. So that is how I have taken it to me. The Qur'an does not say that music is forbidden. It is only one *hadith* that says so. You have to think for yourself, as one African ideological leader says. You have to use your own brain and think for yourself and not just follow uncritically.

Ayşe is fully aware that many Muslims would question her interpre-tation, and this does happen at her work, but it does not seem to bother her. There are as many interpretations of Islam as there are Muslims, as she mentioned during our conversation. She selects the Muslim messages that suit her and with which she can elaborate and heighten a particular self-understanding. Without these more "tolerant" interpretations of Muslim practices she cannot imagine that she could have become a Muslim. Instead, through the personal, selective appropriation of the Muslim faith, her social values acquire a religious meaning and dimension. The Qur'an holds messages that evoke a whole set of already existing ideas about the world, about affection and care for the homeless and poor, equality, and that people are born without sin. The messages in Islam were compatible with already salient ideas. The way Islam was lived in people's everyday life, as well as the powerful sisterhood among women, appealed to Ayşe.

Ayşe's meaning-making of Islam, which is delicately intertwined with and inseparable from her self-making as Muslim, activates two seemingly disparate discourses. Through the *personal model of social justice and solidarity with the Third World* she draws on and reconciles a "Swedish" socialist discourse and an Islamic discourse. Islam confirms and offers religious authoritative legitimacy to her "belief in the good in people, solidarity, justice, and communion."

Becoming a Muslim also brought with it an extended engagement in different activities such as involvement in organizations and associations and visiting schools. The conversion has allowed Ayşe to further explore within a new religious and cultural domain her self-understanding as a woman, engendering an alternative femininity. Through her active participation in women's groups and in international contexts for Muslim women, she has gained a confidence she previously lacked as a "normal Swede." This gendered self-image is negotiated and reinforced by other Muslim women, by the social contexts she takes part in, and by the dominating religious understanding that engages her own sense of worthiness and being a good Muslim.

In contrast to Mariam and Fatimah, Ayşe was not brought up in a religious home and did not internalize a religious belief as a child. Quite the opposite. But she felt attracted to the teaching of Islam with its emphasis on social issues, ideas already of cognitive salience to her. Initially, in her first encounter with Islam it was not religious ideas but social and political ones that appealed to her. Ayşe's personal understanding of Islam takes a shape quite different from, for example, Lisa's and Layla's. Paying attention to the personal appropriation of discourses allows us to see how the interplay between discursive and mental models often leads to the modification of existing cultural and religious forms, as well as to various possible interpretations and variations. In the next chapter I turn to the accounts that elaborate upon gender models while making sense of conversion and the formation of a Muslim identity, pointing to the exploration of alternative versions of femininity.

Chapter Five

Personal Models of Gender

In the previous chapter I discussed how some of the converts' identity formation and conversion is organized and understood through personal models of spirituality and social conscience. In this chapter I continue to analyze the different models of gender that constitute dominant themes through which the converts make sense of themselves as Muslim women and their conversion is at least partially understood. The conversion has triggered reflections regarding gender roles and relations, questioning previous taken-for-granted ideas and exploring new ones. Certain Islamic ideas of womanhood and the nuclear family appeared quite appealing to some of the converts (cf. Sultán 1999). The conversion allows the women not only to explore and rethink themselves as religious selves, but also as gendered selves through different ideas of gender complementarity, equality, attractiveness, desires, and womanhood. The women's identity formation reflects an intricate negotiation and reconciliation of different, sometimes seemingly divergent, discourses.

There are particular dominant discourses or representations of gender that the converts refer to as "Swedish" or "American" and "Muslim,"[1] and which they draw on in their identity-making as Muslims. While the converts elaborate in various idiosyncratic ways on the issue of their roles and obligations, both representations become objects of criticism and scrutiny. By articulating two divergent discourses on gender relations and ideal womanhood, some of the converts produce a critical commentary on both "Swedish"/ "American" and Muslim ideals and practices. In this and the following chapter on the veil, I propose that these commentaries, with personal resonance, involve an emergence of newly sensed femininities.

Below I explore how the interaction between the two personally appropriated representations, "Muslim" and "Swedish"/ "American," take different expressions in each of the women's self-understanding. I begin with Marianne and her interpretation of Islam as well as the compromise she works out between conflicting understandings about how she is supposed to be and act as a woman.

Marianne—Commitment
to Women's Rights

I have met Marianne several times and done three long interviews with her over the course of six years. She is in her late thirties and has been a Muslim for twenty years. During these years she believes she has gone through a process in which she has become more tolerant and open-minded to different interpretations as well as to being able to look at Muslim practices in general with critical eyes without feeling that she always has to defend them.

Marianne grew up in a secular home where religious matters were rarely discussed, and neither she nor her siblings were baptized. Similar to Ayşe, Marianne describes the spirit of the 1960s and 1970s as the "soft-communist period" when religion meant "oppression." For Marianne, the conversion reflects a longing for changes in her life. She was young and found college life "meaningless" and "pathetic." She was motivated by her will and desire to believe in something. "I wanted to believe," she continued. "I *wanted* to feel secure in something." She came into contact with Islam through a close friend who had converted and instead of asking "why?" she asked herself "why not?" As she put it, she could not find any satisfying and good answers to the latter question.

Marianne's self-understanding as an intelligent, educated, and reflective woman is linked to her ability to question different ideas around her and to objectify herself by describing how she assumes people categorize her in different situations. During one interview she openly reflected upon her decision to convert, the difficulties in actually pointing out particular reasons, and the fact that her explanations are most likely reconstructions. Marianne refers her way of thinking to an academic, "postmodern" discourse.

> Islam could be seen as one big human construction or narrative, as silly as all the other ones. Yes, that one has to put everything into perspective [*relativisera*]. But you get to a point, "well, I have to make up my mind for something." I mean, you have in a way decided for it but it could just as well not be true. It is a philosophical problem that if there is nothing, is there then any moral? Is there a meaning to anything? Really, that is where you started when you converted and then you go through it again, that is how it feels. To feel if Islam is valid.

Marianne seems to move away from any kind of essentialism in her own understanding of and approach to Islam. Since many beliefs are "related to culture," she perceives it as a belief, as one among many other "truths." She admits she does not have any rational arguments for why she does not eat pork or why she wears the veil. If somebody asks her why she does not eat pork, she has to say, "I don't know." "Maybe it is a test of my way of life. I don't know why God has decided." She continues,

> This is a belief, I have decided for this, and my clothes are a part of my identity. I do not know why we women should cover ourselves like this. I can find a lot of different explanations or pseudo-explanations why we shouldn't attract other men. But it is a part of my identity and I have accepted it since there is such a consensus among Muslims that a woman should cover her hair. What I find difficult in my life is that since I'm wearing the veil I am questioned. I find it very hard that my intellect is questioned. You can't be intellectual or intelligent and Muslim at the same time. That combination does not exist for most Swedes. In their eyes I have to be somewhat stupid if I, as a Swedish woman, decide to convert to Islam.[2]

The veil and the stereotypes attached to it obviously cause a difficult inner conflict with her own sense of who she is. Because of what she refers to as her "prestige-mindedness" and since she has always been perceived by others as wise and clever, she feels highly sensitive to these kinds of categorizations and misrepresentations of her. It seems to me that she has decided that her Muslim faith is a truth in her life—whether constructed or not, it does not matter. Marianne refers to the philosopher Pascal who said that "it is better to believe than not to believe because if it is true, if there is a God, then it has positive effects and if it is not it does not matter." "And sometimes if I think 'I do not believe' I think that at least I have got a fantastic life because I became a Muslim." When openly reflecting on uncertainties and the meaning of her faith, she told me that if she ever has any doubts about the existence of God she thinks that she *wants* to live as if "He" existed.

As with Ayşe, the conversion to Islam has not only resulted in a religious identity and practice but also in an engagement in social and women's issues. Unlike Mariam, Fatimah, Lisa, and Layla, both Ayşe and Marianne stress other aspects of their Muslim identity as much as the spiritual aspect, if not more, when they reflect upon themselves and their conversion. The most dominant theme in Marianne's account is herself as an "activist" helping Muslim women, discussing different interpretations and approaches to gender issues and involving herself in different women's organizations and publications. In Islam she found a framework in which she could further explore questions on gender and her own identity as a woman. Marianne found that Islam addresses many questions she could relate to and identify with, spurring cognitive recognition. In her personal reading of the Islamic scriptures she not only experiences opportunity to rethink herself as a woman but finds tools to demonstrate Islam from a perspective not often present in public discourses. The challenge itself seems to have been appealing to Marianne.

In her feminist reading of Islamic scriptures, understandings of gender equality and women's rights triggered the conversion and made sense of her Muslim identity. Below, cognitive reconciliation, the inner work of resolution of "new" and "old," signifies an interesting employment and oscillation between two different, seemingly conflicting external representations.

> I have always been interested in women's issues. I believe in much of the same things that I believed in as a Swede, that we have to have equality. Since we in Sweden have an ideal of equality, Islam's gender issues are very problematic to a Swedish woman. The conflicts [for Swedish converts] have always been about women's issues and men's traditions. But then at the same time, Islam also colors you, that you can see Muslim women's criticism of Swedish women. Islam says that it is not equality through sameness [*jämlikhet*] we should strive for but equality in the sense of having the same worth and opportunity [*jämställdhet*]. You do not have to have a job. You could also be at home. I do not put this into practice myself but it implies that working, having a job, is not the most important thing. We who were born during the sixties have experienced our moms to be emancipated and working. But then we have also experienced the negative effect of it like divorces and the like, and that is why it is easy to be critical of that too. In Islam you have the right as a woman to be supported by your husband. But there is nothing in Islam that says that the woman has to be at home. But it puts the emphasis on the family and that we shouldn't be so individualistic. You have to be considerate. Then if this means an absolute obedience or something else, that is what the conflict is about. Traditionally, they say that the woman should obey the man, and that's that. Then we [Muslim women] find ourselves in a kind of dilemma when we say that the man actually has the last word while I myself never accept my husband's last word. Even if he ought to have it, there will be a discussion anyway. You are a human being, right? What it is about is that you have all these ideals of how it should be and in the end somebody has to decide. But in real life it does not work that way.

Marianne voices not only a tension between the "Swedish ideas" she grew up with, internalizing them as a child and as an adolescent, and

Muslim ideas that she appropriated later in life, but also the difficulties in living up to religious principles in the practices of everyday life. When contemplating gender relations, confrontation between two representations comes into play. On one hand is a Swedish one consisting of ideas of gender equality and women's right to work and to have personal projects outside the domain of the home. And on the other hand is a traditional Muslim perspective emphasizing the obligations of the woman as a mother and as a wife, her modesty, and the importance of her presence in the home. From one sentence to another Marianne goes back and forth, comparing, trying out, and confronting the different understandings, and how they could be implemented in her personal world. In the Qur'an she reads that the man has the last word but in her face-to-face interaction with her husband she has difficulties applying this to her own marriage. Marianne has a strong sense of herself as being a sensible and smart woman who questions things around her. She admits the difficulties in giving her husband the last word in arguments, "in real life it does not work that way."

A traditional Muslim message, supported by Qur'anic verses, asserts an ideal role for the woman to be obedient and stresses her obligations in the domain of the home. Family values are strongly highlighted and the woman's role as wife and mother is highly cherished. Marianne embraces these ideas; she thinks that the values promoting the presence of a mother at home are important. At the same time she finds personal satisfaction and intellectual stimulation in her work. She experiences a potential internal conflict between dominant ideas of marital role obligations, on one hand, and of equality and her rights, not only as a woman but as a human being, to satisfy personal interests outside family life, on the other hand (cf. Quinn 1992:116). "You are a human being, right?"

Both messages have cognitive salience and motivational force for Marianne, she wants to be both a good Muslim mother and a woman deriving satisfying results from her work. She has resolved this conflict by working at home as much as possible and not working full-time. The kind of work she has offers her flexibility and a possibility to achieve a compromise between different obligations, conflicting feelings, and needs. She has found her own solution to embrace both of these goals. This is, as Quinn (ibid.) shows in an article about American wives' conflicts in their marriages, an ongoing contest in American society, and, I would like to add, in Swedish society as well. The difference, though, is that the female converts here refer to the conceptions of women's "natural," biologically given, or God-assigned obligations and complementary gender roles as

"Islamic," since they find explicit support for these ideas in the religious scriptures.

The first time I met Marianne, she had a six-month-old baby and was a full-time mother. She and her husband had agreed, before the marriage, that she should finish her studies and get her degree. She expressed, however, a changed attitude toward her goals and future plans to have a professional career. Instead she stressed her duty, supported by a personal desire, to be at home with her children. Six years later when I met her again, the children had grown older and Marianne was working with her own projects outside the home. She still emphasized what "Islam says" about the family as an important thing, and that if she had to choose or if her family was hurt by her work she would leave it. "I do not work full-time and I work at home a lot. I take my family into consideration and try to find a middle way."

The personal meaning she assigns to the Muslim representation of gender relations is infused by previously internalized ideas about equality, in the sense of space for personal projects and divided house labor. Through these ideas that for Marianne compose a "Swedish" model of gender, she interprets and understands the religious message. Marianne stresses that there is nothing in the Qur'an that says anything about a woman not being able to work or have a professional career. She also questions and rejects patriarchal traditional views that the woman should not show sadness or tiredness in front of her husband, that she should serve him and always help his family. With the support of previously learned ideas, and consequently her personal reading of Islam, she condemns the traditional, patriarchal oppression of women and the inequality between women and men. Simultaneously, Marianne criticizes what in her view is the egoistic and individualistic Swedish society, where the woman is more or less obliged to work, where being a mother gets little respect and is seen as having little value. She criticizes "Swedish" or "Western" ideals of appearance that exploit women as half-naked and the lack of morals and family bonds. By condemning these phenomena she points to the Muslim perspective as one offering a better alternative.

In her identity formation and effort to create inner meaning, bringing together different messages, she draws on both representations without fully accepting either of them. Her previous ideas about gender equality could be integrated and attain meaning within her new belief system, reflecting the very work of cognitive reconciliation. Marianne's identity-making as a Swedish Muslim woman and the personal appropriation of two seemingly conflicting and irreconcilable discourses trigger a critical commentary on both Swedish and Muslim messages about how women should be and act. She has engaged and integrated both in a *personal model of gender equality and women's rights*. Making sense of and reconciling the two representations within a personal cognitive framework reflect the ongoing process of identity-making and of integrating a break in life with a coherent sense of self.

This critical commentary reflects what I have earlier referred to as the experience of a special opening to explore alternative ideas. The critical analysis of dominating misogynic messages, by drawing from a new religious belief, creates the possibility to elaborate new gendered self-understanding. The identity formation of Marianne as well as Ayşe involves a new sense of femininity, supported by the meaning assigned to the veil but also as a consequence of increased social engagement with other Muslim women. As in the case of Ayşe, Marianne's everyday life has changed considerably. They are not, as they put it, just any Swedish women anymore, they are Swedish Muslim women with a mission. With their involvement in several women's organizations, both national and international, their commitment to help Muslim immigrant women to integrate into Swedish society while preserving a Muslim identity, holding talks in different public contexts, and getting attention from researchers as well as media, have made them feel important and part of something bigger. They both express a stronger self-confidence as women. Both Marianne and Ayşe described themselves as being very shy persons before becoming Muslim. Since their conversion they have traveled more, got to know a lot of people, and gained an increased social network. They perceive themselves as being part of a larger global "movement" in which older patriarchal views and interpretation of the Qur'an and the *hadiths* are being disputed and refashioned.

During the last few years, several independent Muslim women's organizations, offering different kinds of services and women's activities, have taken form in Sweden. This reflects, as Roald (1999) has shown, an increased interest among Muslim women in taking part in the development of a "Swedish Islam" and an aim to work for integration and a greater mutual understanding between men and women as well as between different cultures and countries.

We have seen from the narrative account of Marianne how different, sometimes conflicting, understandings are negotiated and reconciled within a particular subjective world. Marianne airs both a religious belief in God *and* a postmodern idea of relativism that there may be no "truth" truer than any other. Sometimes she identifies with culturally dominant ideas of what the proper Muslim woman should be. But she also expresses anger and exasperation about the submissive roles of Muslim women. Marianne, like many of the other converts, is a veiled, working professional conscious of her rights and the unequal treatment of women in Islamic as well as in Western countries.

The expression and elaboration of distinct self-representations reflect what has been called dialogic selves (Bakhtin in Holland et al. 1998), shifting selves (Ewing 1990), or contested and contradictory selves (Skinner and Holland 1998). This self, drawing on diverse and conflicting messages, attempts to give self-presentations that will be accepted by others. Marianne's shifting self-representations become rather obvious, even for herself, when she, for instance, describes her different attitudes to her husband's helping hand in two different contexts.

> When we were down there [in her husband's home country] I didn't want him to iron his pants. I told him not to stand there ironing because then they may think that I'm a bad wife. On the other hand, I think it is great if he does it at home [in Sweden] because then they can see that we help each other with the housework. Down there I react "ooh, don't do that!" But here I point out to Swedes how kind and helpful he is.

Diverse and shifting self-representations are, as Ewing (1990) shows in an article about a Pakistani woman's multiple, inconsistent self-understandings, context-dependent and can alter rapidly. In Sweden in front of friends and relatives it becomes important to play down the wifely duties and instead assert her and her husband's fair division of household labor. In his home country, on the other hand, her obligations and role as a wife and her self-representation as a good Muslim woman are accentuated. I have not done "participant observation" but naturally they, like myself, present and negotiate different self-representations in different interactions depending on context and expectations. The projection of incompatible self-presentations shifts from one situation to another, from one moment to another. It can be in the interview with a researcher, at a conference arranged by a Muslim women's organization, in school where they work, at their office, in the post office, or with their parents. This experience of self, constructed through a dialectical movement with the other, is important to acknowledge, but as Ewing (1990) points out, we also have to approach the problem from a psychological perspective.

As the women's accounts attest to, individuals reorganize themselves in response to internal and external changes and processes, but despite this transformation, they still feel a personal wholeness and self-continuity. Ewing calls this experience of personal coherence an "illusion." But, she continues, even if it is an illusion this is a crucial experience we cannot ignore. Ewing asserts that we have to account for this important phenomenon but "without falling into the error of reifying a unitary self" (ibid.:263). I agree with her that this is a central human phenomenon to attend to but, as I will show below, I have reservations to her argument that the experience of personal whole-ness is illusionary. I would also like to assert that it is possible to demonstrate a person's sense of self-coherence without resting on and, as a result, reproducing an idea of a unitary, essential self.

From the quite different postmodernist and poststructuralist perspective, some cultural researchers have depicted individuals in the Western world as fragmented, contested, and split selves (Jameson 1991). With respect to the large portion of cultural messages and disparate discourses around us as well as the flux of everyday experience, this may seem like an accurate description. But, as Strauss argues with the support of interviews with some of these supposedly fragmented Americans, this implies an idea that presup-poses that all of these distinct representations are copied directly and unaltered into people's psyches. This thought, which she calls the fax-model of internalization (1992a, 1997), relies on a simplified image of the relationship between subject and cultural messages. If the individual had acquired the entire spectrum of discourses around her, this would denote a chaotic complexity of cultural representa-tions. Internalizing public messages does not imply copying them in a straightforward way or that these discourses directly construct psy-chological realities (Strauss 1992a:10). The women relate to and express contrary representations, as well as self-presentations, but this does not necessarily have to result in disintegrated identities. So how do people deal with constant change, inconsistency, and con-tradiction? Or, as Ewing puts it, "Why are we not all psychotic?" (Ewing 1990:263)[3]

As I argue, personal models are crucial to look at in our under-standing of the formation of a Muslim identity and the psychological appropriation of Islam. The theory of personal models elaborated here points to how impressions and experiences are filtered through specific salient mental structures. The dynamic models construct a sense of relatedness and wholeness even of seemingly disparate representations. Through the cognitive process of conversion, divergent public messages are internalized, reconciled, and made to one's "own" within a particular personal inner world. In contrast to the case of the Pakistani woman that Ewing proceeds from in her analysis of shifting self-representations, Marianne is highly aware of the changes in self-experience and the acquisition of different representations. I would argue, that the sense of wholeness and coherence is not an illusion, but rather the outcome of the very work of self-making, reflecting the ongoing process of cognitive reconciliation of different representations and self-presentations into a sensed coherence of "who I am" and "my life." There is not necessarily a contradiction between, on one hand, experiencing and acquiring opposing goals and multiple identifications and, on the other hand, experiencing continuity and coherence. A sense of personal wholeness is made possible through cognitive reconciliation that requires an active, self-reflective self.

Aware of changes and tensions between self-representations, the women nevertheless stress a profound feeling: "I'm still the same." Or as Mariam stresses in the very beginning of the book—the feeling of what is "inside" does not really change. The process of cognitive reconciliation, so prominent in the case of conversion, constitutes the very activity of identity formation. Through the *personal model of gender equality and women's rights*, Marianne reconciles diverse messages and representations and in this process she acquires a strong sense of self-coherence, of still being the same, namely a Swedish Muslim woman well aware of her rights.

Marianne understands herself as an intellectual Muslim woman through ideas of equality and rights rather than difference and role obligations. The cognitive salience of these understandings constructs an experienced wholeness despite other self-representations such as the one projected in her husband's home country. In her self-presentation she blends "postmodern" thinking with her religious faith and her commitment to women's rights with rather conservative religious ideas on gender roles and ethics of modesty. As different messages can be integrated, different self-representations such as "a working professional woman" and "a modestly dressed Muslim wife" can be expressed and reconciled through personal models.

Cecilia—From "Party Girl" to Modestly Dressed Muslim

I had the kinds of parents who brought me up as well as my sisters and brothers to always question everything. Especially things you can read in the newspaper and what is said on TV and what authorities say. If it sounds idiotic question it! . . . A major idea in Islam is that you should study and learn and question things until you find a satisfying answer. And that fits my mentality much better since I'm such a person asking "why that?" [Cecilia alters her voice somewhat, sounding like a little girl]. I'm one of those precocious persons, even when I was small I asked "why that?" All the time. And if they didn't give me any reasonable answer I refused to listen to it . . . And that is the biggest difference between Christianity and Islam. If the Christian priest tells you to jump you shouldn't ask "why?" but "how high?" In Islam if somebody tells you to jump the first thing you should ask is "why should I jump?" And that fits me better. Why should I jump just because someone tells me to, if there is no logical reason? I don't buy the idea that you should do something just because someone snaps his or her finger. That doesn't exist to me.

Cecilia proudly presents herself as a "cocky" and contentious woman, always questioning the taken-for-granted and dogmatic attitudes. She is one of the youngest of the women I have interviewed, and not even a year had passed between her official conversion and the first time I met her. Unlike some of the other converts, Cecilia did not present a "readymade" self-narrative. She has not had much time to work out a well-defined conversion narrative with a leading, salient theme or idea. It takes time and experience to reorganize a self-understanding and to form a harmonious biographical presentation. Rather, the interviews offered means of trying out alternative ways to approach and make sense of the conversion, a major change in her life. Everything is still new and she does not really know the possible outcomes of her decision and she has just gotten accustomed to implementing her Muslim faith in everyday practices.

Cecilia gave me the impression of being a lively and intense woman, very verbal, talking fast, jumping from one subject to another. While telling me about her background and the very different kind of life she lived before she became a Muslim, she conveyed a portrait of herself as a strong-minded and independent young woman. In a short and compressed account she gives a sweeping description of her life before the conversion.

I have done just what everybody else did: attended kindergarten and daycare, school and high school . . . And when I was sixteen and started high school I moved in with my father. And after that I didn't know what to do, so I worked like crazy one summer and saved a lot of money. And my father said, "go to Israel." And I thought all right and then I went to Israel. I was supposed to be gone for eight months but didn't come back until three years later. And I came back and that

wasn't too nice . . . it's cold here. And it is warm in Israel [laughter in
her voice]. Then I worked at a hotel for two and half years and then I
went abroad again. I worked on cruise liners as a receptionist, as a
captain's secretary, and then I came back to Sweden again.

Anna: Where did you go?

Cecilia: It was in South America, in the Caribbean, and the
Mediterranean. And I came back since I was tired of aimlessly
roving around. You get to this point when you feel "is this what I
want to be doing?" You get so lazy when working on these cruisers,
you don't need to do anything except put on make-up and fix your
hair and go to the office in the morning. Then someone polished my
shoes, washed my clothes, and made my bed. Yes, everything. So
I got a little absent from reality.

After a couple of years working on different cruises Cecilia
returned to Sweden and started to work at a nursing home, still traveling a lot during her vacations. She was then around twenty-eight
years old. At her work she met some practicing Muslims and she
started to ask them a lot, ending up reading literature about Islam,
first out of pure interest, but then later with more and more personal
involvement.

Cecilia grew up in a religious family. Her grandmother is Catholic
and Cecilia was as a young teenager a member of an organization for
young Catholics. But she had problems with some major ideas such as
"the Trinity, that God is three and then suddenly one, and original sin."
"How can someone invent such a hopeless idea that a baby is born with
sin?" Cecilia asks herself during our conversation. These are Christian
dogmas that many of the converts brought up as difficult to understand
(cf. Poston 1992). In Islam Cecilia found an appealing logic.

The logic is very simple. There is only one God; there is nobody
besides God. There are no priests who stand between God and me but
I have straight communication with God by myself. It should be said:
I have been religious since I was a small kid. I have always believed in
God. But I have always had difficulties with the idea that Jesus had to
die as a poor thing on a cross so that I should have my sins forgiven,
which were already there before I was born. That doesn't make sense.

When I asked her how long she studied Islam and about her
Muslim friends she replied,

It wasn't really close friends; it was simply acquaintances. But since I am
who I am I bombard whoever with questions, harass whoever . . . if

I want something I approach it as a steam-engine, that is the way it is. Or I should rather say steamroller.

Anna: It works with steam-engine too.

Cecilia: Well, but they don't approach as well. Steamroller is better— they crush everything in their way. It took a pretty long time for me . . . this process of becoming Muslim *myself*. I couldn't call myself a Muslim until two years ago. I didn't convert officially until the beginning of April this year. So it took a pretty long time before I decided. I also felt that it was of totally vital importance. It is not something you do in a twinkling and then tomorrow I'm a Buddhist.

[. . .]

I believe everybody, no matter if one chooses a new belief or a new lifestyle or a new career or whatever . . . there comes a day when you have to take that step. Maybe it is more visible when it concerns a religious belief than choice of career but in some ways it is the same thing . . . you have to leave something secure and familiar and take a step into something partly unknown. And you don't know how it will be.

Cecilia was on her way to Islam and no one could stop her. By using the metaphor of a "steamroller," crushing everything in its way, she captures her own determination and readiness to take a step toward a new direction in life. Like the other women, she also portrays the conversion as a long process but also as a result of reaching a stage in life facing a need for something new. Becoming a Muslim is described here as leaving a secure and known presence for an unknown future. The "secure and familiar" is experienced as restrictive, not allowing for changes, holding back the self, while the "unknown" presents itself as a tempting and open-ended prospect offering indefinite possibilities. Cecilia was ready for a change. To make sure she was doing the right thing, she tried it out, testing what it meant to implement the belief in her everyday life, during one year before the actual ceremony.

It was simply the case that I felt that I had finished one phase in my life. Now it is time for the new. I guess one can say that I was rather much of a party girl. I was going out quite some and I thought that now it has to be enough. Now I have tried this. I was out partying four, five times a week.

Anna: Wow, how did you manage it?

Cecilia: Yes, I ask myself sometimes. How on earth could I go on and most of all, how could I afford it? Okay, there are always some stupid men who pay for you [laughs]. I had got to that point that now

I have done this, now there is nothing left to do. Now I have to deal
with what is important. Because I felt that I had traveled a lot, met
a lot of people, I had tried a lot of new things. I had have time to
party. Now I felt it was time for next phase in my life.

Anna: Your life partying, was that something you did until the time
you converted?

Cecilia: Well, until a year before I converted. The last year [before
converting], you could say I lived like a nun. Well, not in the sense
that I went to a convent or went and hid myself, but I simply just
stopped living like that. I think it was a practical test to see if I was
able to live without it. And I discovered that it was great! Wow, to
be home one evening at nine instead of coming home at nine in the
morning.

The meaning Cecilia assigns to becoming Muslim concerns her
own gendered self-understanding and previous experiences as a
woman. Like Marianne, Cecilia feels very concerned with women's
issues but approached the problem from a different perspective. She
identifies herself as a feminist but bases her discussion on explicit
arguments about men and women's biological differences. Cecilia
has always been engaged in women's rights and she finds support for
her ideas in the Qur'an—she has "the world's greatest feminist" on
her side, Prophet Mohammad. For Cecilia, equality means that both
men and women have certain rights and obligations. These rights
may not be the same, "because men and women are *not* the same,"
but both should show respect for the other and value the differences.

She presents a *personal model of gender complementarity* comprising ideas on gender differences, a biology-based feminism, and on
what constitutes a good marriage. Both understandings, drawn from
the image of a complementary union, have emotional resonance to
Cecilia and are connected to significant life experiences. The heart of
the model lies in the idea of psychological and biological differences
and compatibility. Cecilia criticizes ideas that women have to act like
men in order to be respected and regarded as equal to men. She consid-
ers these to be particularly "Swedish" attitudes. Instead she engages
a view where gender qualities are biologically given. From her perspective, women tend to think more in terms of "we" while men fight
their way through. In Cecilia's view, women in the Swedish society
are forced to behave like men to get and keep power. She adds, "This
is wrong, women must be able to continue to be women."

It is particularly in reaction to Christian understandings that she
grew up with that she shapes and organizes her ideas about the relationship between men and women today. Cecilia finds support for

her feminist ideas not only in the Qur'an and other Islamic scriptures but also through negation of certain interpretations of Biblical messages that she heard when growing up.

> I have a rather feminist view on things. I refuse to see the woman as a second-class citizen. I refuse to see why a woman should be worth less than a man . . . I like Islam's emphasis on the rights and obligations the woman and the man have. This is very, very important. My rights as a woman are very important to me. I would never obliterate myself for a man, never. And I really dislike a religion which says that the woman should keep quiet within the church. That doesn't work for me. A religion that tells me to shut up because I'm a woman, I don't like that. A religion that says that it is okay that a man hits me to teach me the Lord's good discipline, I don't like that. It is the words of Paul or Luther. The man should be the master of the woman on the earth just as Jesus is the master of the church. But if another religion tells me that I'm my husband's equal in religious matters and have the same right to express myself in religious matters as a man, and that I have rights in society and in the marriage, maybe not in the same way as a man, but still I have explicit rights, that is a religion that fits me.

In her identity-making as a Muslim woman, concerned and highly conscious about her rights, she refers to and engages religious representations. She reads a lot and in her arguments with her Muslim husband, she looks up the standpoint of highly respected Muslim scholars on certain issues. Her husband, who was born in a Muslim country, has, according to Cecilia, ideas based more on traditional customs than religious teaching. While he has internalized the religion as a part of everyday living, affected by cultural and traditional views, and has never really read the Qur'an, Cecilia reads it with very different eyes. This has been the source of some conflicts. When I asked her what issues they argue about, she replied that it often concerns very private matters but also issues such as music, whether it is allowed or not to play music.

Cecilia makes sure no one, including me interviewing her, could possibly view and categorize her as the stereotypical, subordinate, defenseless Muslim woman. She has always been a "cocky" woman and why would the conversion to Islam bring about any changes to this image? She converted, as she puts it, to Islam, not to any patriarchal traditional values.

> I'm very good at talking down persons. And I'm also, as I said earlier, very stubborn. If there is anything I wonder about I will look and

search through thousands of books until I find the answer. And if my
husband says anything different I throw the book in his face saying:
"NO, this is how it is!" Then he has to accept. That he has learned.
Anna: So you use the Qur'an to support your arguments?
Cecilia: Oh yes! Oh yes, I don't stop at anything when it is about those
    kinds of things. That is the way it is. And he can be angry with me
    during the time but afterwards he says, "It is great that you try to
    look things up" . . . I am who I am and I don't accept whatever.
    Instead I look things up myself. And even if he sometimes thinks it is
    hard, he thinks it is good that I look things up. There are many things
    he doesn't think of himself but just does them until I say it is this and
    that. And then he looks in the book and says: "Yes, you are right."

Cecilia invokes a voice from childhood, asking "why that?"—a voice
that composes a considerable part of her self-understanding, questioning and resisting a traditional Muslim representation that could
silence her, a representation that states that the man has the last word.

Soon after her official conversion her closest Muslim friend asked
her if she was interested in meeting a male friend of hers and her husband's. A meeting was arranged as per Muslim rules with a third party
present. Two weeks later they married. The rumors that a Swedish
woman had converted to Islam spread rapidly among single Muslim
men in the city. Women who have converted to Islam are more attractive than nonreligious, Swedish women. "You have ten men per
woman," Cecilia says humorously. Most often the decision to marry a
Muslim follows automatically the decision to convert to Islam. The
encounter with her husband-to-be was naturally perceived at first as
very different and strange, but at the same time she found it rather
straightforward and open. Comparing it with the encounters at discos,
and her previous dating experiences, she describes it as follows:

Of course it is kind of a rather weird situation. If you put yourself
above observing the situation from the outside, you kind of giggle for
yourself. But at the same time, in a strange way, it is also a relief,
because he knows what I want and I know what he wants. There are
few mistakes or misunderstandings since it is direct . . . Both are very
aware of what it is all about. Very different from going to a disco and
meeting a guy not really knowing what he wants. Here we both know
that we want to get married so the questions are rather raw, right on.
"Okay, and how will it be with the cleaning? Who will take care of
this and that?"
Anna: That is how it works?
Cecilia: Yes, you have to ask. That is not something you can talk about
    afterwards. You have to talk about all these sorts of things to see if
    you match each other and to see if you share similar approaches and

interests. Yes, what wishes one has and what one wants from a marriage. He may want to have a woman who is at home every day. And I want to continue to study at the university, and then that wouldn't work.

Why does a Swedish woman agree on meeting and marrying a man on these very different conditions? Cecilia had been dating other men before she started to practice Islam. Maybe she was tired of the "game" at discos? Maybe she had gained a disillusioned view of dating and love? Or as she pointed out in a sarcastic way: "I don't believe in the kind of love of harps and violins. It doesn't work." The Muslim way of dating offered a different way of interacting with men. The frankness appealed to her: this is who I am and what I want, who are you and what do you desire from a marriage? Cecilia describes her relationship with her husband as based on friendship as well as love that developed later in the course of their life together. She liked him from the beginning, and that knowledge was enough for her to make the decision to marry him. She knew there was a possibility that she could fall in love with him.

Though she still holds onto some of her earlier important values, Cecilia has gone through considerable changes in lifestyle. Together with her newly married husband she lives a very different life today from the Cecilia who just a couple of years earlier was a young, single woman, searching for adventures, traveling the world. Her self-presentation reflects a person who was ready for a new kind of life and desired new ways to define herself and her femininity. And her life has changed. At the time of the interviews she was pregnant with her first child, living a married life, barely meeting any of the friends she used to go out with, and practicing her Muslim belief.

Cecilia's newly acquired femininity, the respect and modesty of a Muslim woman, put an end to dyeing her hair and wearing revealing clothes. From being a party girl in jeans and tight tops, from having been perceived as a "sex object" when walking in town, she is today a veiled woman, who keeps men's gazes away from herself. She has appropriated different ideas of sexuality and female behavior and appearance, through which she redefines her femaleness. Walking around in modest Muslim clothing and the veil, Cecilia experiences the changed judgment and image of her.

This may sound very, how can I say it, complacent, but I have always been slim and I have always had a remarkably good-looking body. And that has been a basis for many people in forming a judgment of me. I was only a walking body. And even if you wear jeans and a

T-shirt people can tell how you look. With the veil and so-called decent clothes I get judged in a whole different way. Since people can't see my body they can't think "wow, what a waist, what a sexy butt and what legs!" Instead people have to look into my eyes. Contrary to what most women believe, it is not so much fun to be so-called "sexy." It is not nice when people forget that you actually can have an IQ of 120. Because, in their eyes, if you are sexy you are also stupid. Then I get so damn pissed [whispering] . . . I guess I'm still rather vain in the sense that I try to match colors and patterns and so. It should look nice and decent, not fashion-wise, but fairly modern. When my husband and I are home I fix my hair and put some make-up on. So I still do that but not in the same way as before, outwardly. Now it is something I do privately. I can still stand half an hour thinking what veil would fit best with what clothes [laughs]. I have nineteen veils at home in different colors and patterns. I don't think there is an excuse to look silly just because one is Muslim. I believe one should look neat and rather well dressed anyway.

Cecilia knows herself as a sincere and smart woman and therefore it is hard for her to be approached in another way. She expresses her personal experience of being good-looking and sexy, and for this reason having difficulties being recognized and approached as an intelligent woman. A Muslim identity, veiled and modest, was experienced as a compelling alternative to being "only a walking body," having people form their opinion about her based on appearance. Today Cecilia is no longer judged as a "sexy and stupid" woman but has on the other hand become an object of other kinds of judgments and stigmatization, once again unwanted and misrepresented. Following a Muslim tradition, Cecilia expresses her sexuality within the home, for no other man than her husband.

A critical commentary is verbalized against both Western female ideals and practices such as exuding and revealing sexuality in public, which Cecilia has herself experienced, and traditional Muslim ideals regarding the quiet and submissive role of the woman. Like Ayşe and Marianne, Cecilia engages both Swedish and Islamic representations, but her gendered subjectivity, her new femininity, as well as her critical commentary take another form. In the veil she finds the protection and security from the life she wanted to leave, but also the integrity and respect she missed in her previous lifestyle. In her appropriation of Islam, prevalent ideas about biological differences between men and women and gender complementarity, entailing equality, are reinforced and strongly linked to her self-understanding and her own desire of unwavering companionship. Through the process of conversion she has reworked her feminist ideas, integrating Islamic ideas on gender roles. These gain cognitive salience since they address particular personal experiences and quests.

For Cecilia, the conversion to Islam is not only a conversion in religious terms; it is also a transformation, reflecting a longing for another kind of life and change of self-identity. "I felt it was time for the next phase in my life." Becoming a Muslim then implied not only exploring a new sense of religious self but also a gendered self, rethinking her identity as a woman adapting to things such as the traditional regulation of dating and new ways of displaying her female identity in public.

All of the women I interviewed married a Muslim man, mostly after their conversion. Besides the aspect of a romantic relationship and companionship, I believe that the marriages also reflect a profound dimension of the conversion itself, namely a desire to be "discovered" and recognized as a Muslim woman by a male spouse—a dialogue between the inner formation of self and a social recognition of the transformed self. The convert's gendered and religious identity can in that way be approved and confirmed.

Mariam—"A Different Career Choice"

As I have shown, two dominating gender discourses were repeatedly voiced in the converts' accounts, namely the discourse of female marital obligations as mother and wife and the discourse of individual independence and the pursuit of professional career. While the women are more or less required to relate to these, their appropriation and understanding of them take diverse personal resonance.

Compared to Marianne, Mariam does not work out a compromise between the two. Instead, she felt she had to choose. As we have seen in the previous presentation of Mariam, she learned both from her school and from home that having a meaningful professional career was a primary goal to achieve. She grew up learning that being a mother and wife was not enough. The school she went to was, for that time, very progressive with an explicit political goal that the students, who were all female, should work toward a professional career. In the second interview I asked her to talk more about this outspoken message:

> *Anna:* In the last interview you kind of brought up different cultural images and views of women and their roles. You told me that you were taught to do things, you were raised to do things, never marry or have children, and then you said you did, as you called it, "the

naughty thing for a woman to do." Do you think you could develop this further?

Mariam: Well, I don't know, I don't think there is a general American role. It just happened to be a small part of America that I grew up in. It was a very goal-oriented, cause it was an all-girl school and it was started in order to liberate and educate women. And the goal of it was to bring educated women into the workforce. You know, it had a political aim. It was implicit in the education that . . . not that you wouldn't get married and have children, but that that wouldn't become your main purpose or main focus in your life. The purpose of your life should be some meaningful work.

Anna: And this was a strong message?

Mariam: Oh, very. I mean, it was expected. We used to say: I'm going to get married and have children, *but* I'm going to do this and that. It was . . . you never said that you wanted to get married and you never said that you wanted to be a mother and have children. It was always what we were going to *do*. And I think that most of the girls that I went to school with have done that. They have had careers and if they could fit it in they would have one or two children.

Anna: Do you mean then that your view of yourself as a woman has changed?

Mariam: Well, I think that my mother thought . . . she hadn't gone to college and her father had raised her that she was going to take care of him, and stay home. And she would rebel against that and she didn't want that for me. The message was: don't ever get dependent on a man. Be independent. And it kind of took me by surprise, I mean, I switched very quickly in my attitude after spending time in the field. And, but I think, the way I understood it was: if you are going to do something you are going to do it well. I mean, that was also part of the message. And if I was going to have children and be a mother and a wife I was going to do it well. And I didn't see how I could do it well and do something else. It would either take first place or it wouldn't. And so it was a career *choice*, I feel. I mean [laughs] as much as anything is a choice. I felt that I had a choice. I didn't get married until I was thirty, thirty-one. And I didn't have a child so I felt it was a choice. And then I wanted to do it whole-heartedly, you know, the best I could. That's why I home-schooled rather then sent them to school. I mean, I chose it as a career. I don't think that was acceptable, nobody really understood it. When I was growing up it wasn't one of the possibilities of a career. It stood as opposed to a career. Your career couldn't be being a mother. It wasn't one of the choices. And then trying to grow up again in the field, I understood it then as one of my choices. And I chose it.

Anna: You mean that you saw another way of living and—

Mariam: Yes, another opportunity. And I was sick of watching other people live, and I also wanted a life. I didn't really want to keep on watching other people's life [laughs]. So I felt that I was ready.

At a very young age Mariam internalized a powerful message of having a professional career and becoming something. For many years this model had a motivating force for her, manifesting strong social expectations. However, during her fieldwork in northern Africa, something radical happened to her. She had some emotion-laden religious experiences and encountered other compelling gender roles and obligations and saw how they were played out in everyday life. In the field she "switched very quickly" in her attitudes toward objectives in life. Mariam felt that she had other choices, and whatever she chose, she was going to do it "wholeheartedly." That was also a part of the internalized message: whatever you do, do it well. Mariam could not see how it was possible both to pursue a career and to have a Muslim family and still succeed well in both. While learning the culture and the religious practice of the people she was studying, Mariam also experienced an opportunity to view herself as a woman from another perspective. When she later met her husband in the United States and they decided to have a Muslim family, she dropped her Ph.D. She did, in her own words, "that naughty thing for a woman to do. I did the proverbial thing, got married and got children."

Mariam, who was about to convert, perceived the alternative gender roles as legitimate and quite appealing, partly because they were connected to and authorized by the religious belief she had embraced. We know from earlier discussions that her identity formation is mainly organized around a *model of spiritual understanding and connection*, intimately linked to key childhood memories. This personal model has cognitive salience to Mariam, it both presents appealing ideals and explains emotionally salient life experiences. In the previous chapter we see how she reconciled her earlier religious ideas with Islamic and Sufi ideas. Likewise, through the process of cognitive reconciliation, the competing messages of being a mother and wife, supported by a religious belief, and of pursuing professional career were all inspected and given new meaning. The message of being independent of a man as a working professional was reconsidered after the conversion.

Appropriating a Muslim faith not only offered compelling spiritual ideas but also powerful understandings of her role as a woman and desirable goals in life. Drawing from it, Mariam challenges the role previously expected of her as a professional. She elaborates a critical voice against cultural expectations on women, resulting in a tension between marital obligations and the pursuit of a professional career. Through the process of becoming Muslim, both in the field and later within a Sufi community in the United States, the cultural message of individual independence and of having a professional career had less and less cognitive salience and motivational force. The *personal model of spiritual understanding and connection*, now encompassing Muslim ideas of the modest role of the woman, has gained stronger directive force than that of the cultural ideas of having a successful career (cf. Strauss 1992b). It permeates her ideas and feelings about gender relations offering appealing ideals, addressing personal desires of motherhood.

Hannah—The Yearning of Strong Family Bonds

Like the other two American-born women, Mariam and Fatimah, Hannah was brought up in a Christian family and religious ideas have always been of importance to her. I discuss her self-understanding and formation of a Muslim identity in this chapter, however, since the ideas of the significance of strong family bonds and values stand out as the most prevailing in her account.

Hannah is of African American descent, born in Mississippi and brought up in Indiana. Her grandfather was a minister and her mother is "this Christian who really believes in going to church." Accordingly, she grew up in a religious environment, attending church every Sunday with family and relatives. Early on, however, she felt, as she expressed it, "out of place."

> When I was growing up, all of my cousins that were in my age, they were really into going to church and I went to church too because I was a kid and I had to go to church. I didn't have a choice but I didn't have that interest like they did, so I thought that was weird the way I felt. Even when I was twelve and thirteen.

There are some key childhood memories that have had a great impact on her self-understanding and the choices she has made in life. As a child Hannah witnessed a lot of betrayal and cheating between married couples within her religious community, and at a very young age she made up her mind on what was the most important and meaningful thing to her. Above all she wanted to get married to a truthful and caring husband and have a family with strong ties. These were values she found being practiced among Muslims and stressed in Islam, which she came across as a young teenager. In her initial encounter with the religion these ideas were the most compelling to her. The conversion was spurred by a strong personal quest for strong, reliable family connections. During the first interview Hannah talked primarily about family matters and family relationships.

> Well, to me, being a Muslim, being married and having family, trying to raise my children is important. And to me when I read a lot of things about how the Muslim man looks on the woman . . . The man is the head of the household, he is the leader. When I read that it really got my interest. I hadn't had a relationship with a man . . . *but I had seen things.* I have seen my family members who are married but have boyfriends. You know the husbands have women all over the town. I had aunts that had maybe ten kids. And the man that they had kids with was married. They all go to church together. She got these kids and the other wife has kids but they don't know they have a father in common. You know they are sneaking around doing this. I looked at that and how the minister looked at that and reacted. He was a womanizer. The men I knew who were in the church always had two women even if they were married. And the wife didn't know. They didn't really take care of their families. And when I met some Muslim women who were married, their men held them very highly and took care of them. The relation I wanted was a husband who took care of me . . . and thinks of me as important, you know.
>
> *Anna:* You mean that you had seen things in Christian families that you didn't want.
>
> *Hannah:* I knew it wasn't right. But when you read the Bible it says differently. They weren't living it. And when I was reading about Islam how you really lived the religion . . . that really interested me. [. . .]
>
> *Anna:* So you saw a lot of betrayal in your childhood, which made you think that you didn't want that.
>
> *Hannah:* Right. My aunt had eleven kids. She was married and had four kids with her husband while she got the other ones with another guy to which she wasn't married. I have another aunt who has six kids and she is not married.
>
> *Anna:* How is this viewed in your family?
>
> *Hannah:* Well, my grandfather doesn't like it. But they still tell you that they are Christians . . . there are some Christians that are really into their religion, hard-core Christians, and they go by the book. When I look back at it no one was really practicing the religion. They said they were Christians and they went to church on Sundays but when they went home for the rest of the day they were living in sin.

Unfortunately, and paradoxically, the price she had to pay for converting to Islam was broken family ties with her sisters and brothers.

Hannah feels that her family denies her Muslim identity and shows little understanding of her choice in life.

> My thing is family. I feel that family should stick together no matter what. I mean, they should be close. I think that family should be close. That is one thing that I try to hold on to. Although, like I said, I feel distant to my family because I live far away. I feel that family ties have got away from people. I feel it is sad that me and my sisters and brothers don't communicate. I have called them, I don't call them often, but they never return my calls. You know what I'm saying? It is hard for me to be the only one trying to hold on to this.

Hannah's identity as an "African American Muslim" woman is organized around certain compelling understandings of traditional family values, stressed by other religious groups as well. In Islam and the way the religion was practiced by Muslims, she recognized a meaningful message about the importance of family, speaking to personal ideas and dreams. She felt a powerful emotional appeal to the family ideal and the messages about complementary gender responsibilities, particularly to how the man should treat and take care of his woman, meeting her own desires for an intimate, faithful, and secure companionship. It is through a *personal model of family values*, linked to childhood experiences of betrayal and unfaithfulness, that she feels her commitment to Islam and perceives herself as a Muslim woman, wife, and mother. The words in the Qur'an that "paradise is at your mother's feet," as well as the explicit message that the man has the financial responsibility and obligation to support her, are intimately linked to her self-understanding as a Muslim woman. It offers her a feeling of protection from being hurt and deceived, and a sense of being guarded from divorce, which her own parents went through when she was a young girl.

In high school she got to know a Muslim girl who introduced her to some other Muslims. This was the beginning of a period of reading and thinking about Islam. After she finished high school she met her future husband through her cousin's husband. He had converted to Islam some years earlier and he began to teach Hannah "the ethics of religion." At the same time she started to think about marriage and how she wanted her family to be. He was six years older and since she was only seventeen both her parents and her parents-in-law were against the marriage. But they married at an early stage in their relationship after traditional Muslim directive and at the age of eighteen she officially converted to Islam. During the interview it became important to Hannah to assert that so far her aims and dreams have been fulfilled. In a twenty-eight-year-long marriage she is not only a successful and proud mother, but also a grandmother.

> So it has been twenty-eight years. It has been so many years it is easy to lose track. We have never got separated . . . and we have five children, as I said. My daughter and son are married and have kids so we have also grandchildren.

Chapter Six

The Veil and Alternative Femininities

The symbolism and meaning of the veil comprise all the themes discussed in previous chapters through the different personal models, touching upon spiritual, political, and gender issues. The veil[1] is a widely disputed and charged symbol with diverse political, social, cultural, and personal meaning depending on geographical and historical context, in Muslim as well as non-Muslim countries. The issue of its purpose is quite complex and emotional, and the interpretations vary greatly. Studies have shown that the meaning of the veil is quite different from one country to another, from one situation to another (El Guindi 1999). The use of the veil can mark certain class affiliations and economic privileges, political struggle, protest, and opposition and reflect larger social and political transformations.

In Western discourse the "veil" is politically charged with connotations of the inferior "other," suggesting the subordination and inferiority of the Muslim woman. Also within Western traditional feminism the veil has been perceived as a controversial and provocative symbol of patriarchal oppression of women. But as Bulbeck (1998), among others, has pointed out, this is a one-sided perspective. Western feminism has to reorient itself and enter into dialogue with other ways of defining gender roles and freedom. El Guindi (1999) has also criticized women's studies for studying the veil only in the narrow domain of gender rather than as a differentiated and variable practice, implying different meanings depending on context. First and foremost, the donning of the veil reflects the significance of belief for the woman. It is perceived as an Islamic obligation, and a manifestation of one's religious commitment and belonging. In this chapter I focus in particular on Fatimah's experience of the veil, but before that I give a brief overview of some general ideas and arguments about the *hijab*.

Gender Segregation and Protest

Scholars have shown how the recent modern veiling movement in countries such as Egypt has made it easier for women to work outside the home and still maintain societal respect (Hoodfar 1991, MacLeod 1991). In Java the growing inclination among women to veil reflects a challenge to local traditions and a protest against the government and its politics (Brenner 1996). In the first example, the veil seems to function as a protection for the women, allowing more freedom when moving within the public sphere, which in some Muslim societies is considered a male space (Mernissi 1975).

The Muslim feminist and sociologist Fatima Mernissi has analyzed Islamic patriarchal ideas about the difference and segregation between men and women and female sexuality. She describes the veil, in Arabic *hijab* (note: *hijab* can mean "veil" or "curtain," something that separates, depending on context), as three-dimensional. It has a visual dimension that implies hiding something from sight. The second dimension is spatial, meaning "to separate, to mark a border, to establish a threshold" (1991:93). The third dimension is ethical, pointing to the realm of the forbidden. She continues, "A space hidden by a *hijab* is a forbidden space" (ibid.:93). Mernissi questions the prevailing interpretation and use of the veil as something separating women from men by referring to verse 53 of *sura* 33 about the "descent of the *hijab*." The *hijab*, here referring to a curtain, originally "descended" through a revelation to the Prophet from God to separate a male visitor from the Prophet and his newly wedded wife. The Prophet Mohammad did not want to be disturbed and he therefore marked the space between himself and his companion with a curtain. As Mernissi points out, "A relatively minor incident—after an evening meal some guests delay their departure longer than they should—provokes a response so fundamental as the splitting of Muslim space into two universes—the interior universe (the household) and the exterior universe (public space)" (ibid.:100). The verse presents, in her view, a division between public and private, between the profane and sacred, but was later turned into an understanding of gender segregation.

In an earlier book, *Beyond the Veil*, Mernissi explores the issue of male-female dynamics in Islam. Among other problems, she discusses what she calls the implicit theory of female sexuality, reflecting an interpretation of the woman as active and embodying *qaid*, namely, power, "the most destructive element in the Muslim social order" (1975:5). This model acknowledges the importance of sexual satisfaction for both men and women to maintain social order as well as the fear of a woman's self-determination and her sexual power over a man. In comparison to the Western Christian tradition and experience, female sexuality is viewed as active rather than passive (see also Mernissi 2001). Moreover, sexuality itself is not under attack or degraded as something wrong and bad but is instead encouraged, as long as it is channeled within the marriage. Sexual desire should not be suppressed but instead harnessed in the right way. It could then serve Muslim order. However, what is denounced, in Mernissi's view, is the woman.

In this Muslim discourse the woman embodies destruction and disorder. She is *fitna*—chaos provoked by sexual disorder. Referring to Imam Ghazali's work on female sexuality in Islam, written in the eleventh century, she presents one, in her mind, prevailing Muslim idea that "women must be controlled to prevent men from being distracted from their social and religious duties. Society can only survive by creating institutions which foster male dominance through sexual segregation" (1975:4). In this context, *hijab* becomes an expression of and a means for this separation and protection. These ideas have composed a dominant Islamic representation of gender relations and female sexuality.

However, as many recent studies have shown, the modern veiling, partly as a component of the global Islamic movements, represents new and different goals and ideas. Also the Islamic attitudes toward women and their role are about to change in European societies due to the Western emphasis on gender equality, the integration of educated second- and third-generation Muslims, and globalization (Roald 2001:300f.).

In her interesting study about veiling among young women in Java, Brenner (1996) understands it as the means of self-transformation as well as larger processes of social change. Donning the veil is then both a conscious remaking of self through Islamic discipline and a step in remaking society. In the particular context of Indonesia, the recent movement of veiling represents a protest against the government, local and past traditions, and Western models of modernity. A main point of hers is that it reflects an "alternative modernity" without secularism, materialism, and social alienation (p. 678).

Brenner's analysis of veiling rests on theoretical assumptions of individual agency. In contrast to a perspective proceeding from Foucault, understanding veiling as "a certain technology of power over the body," Brenner stresses that "to reduce veiling to an effect of totalizing forms of power on individuals elides both individual agency and the symbolic role of veiling in processes of self- and social production" (p. 709). If we disregard individual agency, we will miss not only the reconstruction of self through veiling but also the resistance the Javanese women ventilate to both "traditional" visions of womanhood and the gender ideologies upheld by the regime and mass media. She continues, "if one wishes to look at veiling as an inscription of power relations on women's bodies, as many people have, then one must also recognize the potential of veiling for destabilizing or refiguring those relations of power" (p. 710).

In a similar theoretical vein, Mahoney and Yngvesson (1992) argue that we have to integrate analysis of motivation and agency of the individual when documenting the choices and situations of women. What makes these Western women choose to veil? What does the veil mean to them? With these questions I pay attention to the desires and requests expressed by the women and recognize their capacity to engage and make meanings in their interaction with others, instead of seeing them as mere "victims" of culturally constructed subordinate positions or their bodies as passive recipients of power. Attending to personal experiences and desires and the meaning assigned to cultural representations and symbols, we might gain a better understanding of why they conform or resist certain ideas and relationships of power.

For example, what does it mean when Fatimah says that the veil is *a part of herself*? Fatimah does not only identify with a cultural and religious symbol, she also invests emotions in the piece of cloth that covers her hair. She gives it personal life. The veil is a symbol infused with particular personal meaning. In understanding the commitment to and motivation for veiling, the veil can be approached as a *personal symbol* (Obeyesekere 1981), namely it has meaning both on a cultural and on a psychological level. Veiling mirrors not only the women's engagement in cultural meaning and social practices but also personal desires, concerns, and certain life circumstances. The veil signals the women's profound religious faith and commitment. It not only symbolizes some shared Muslim ideas and a sense of belonging but also signifies an inner personal world. The meaning of the veil lies in the biographical particularities of the convert and her experiences of interaction with others. Hence, the *hijab* is psychologically meaningful. Veiling, as a significant part of the conversion, engages and resolves personal dilemmas and concerns (and causes others). Departing from the reflections of Fatimah and her emotional investment in the veil, I demonstrate how it operates as a personal symbol and what meaning the veil can have in a particular personal world.

Fatimah—The Veil, "My Home, My Modesty, My Privacy"

Veiling is strongly linked to the formation of a female Muslim identity, and to ideas about gender relations and modesty. All of the women, with the exception of Zarah, were at the time of the interview married to Muslim men (however, some of the husbands were not practicing Muslims). Even if the conversion itself was not a direct consequence of marrying a Muslim man, some of the women, such as Layla, Zarah, and Ayşe, were introduced to Islam through their husbands. Many of the women emphasized the exploration of Islam with their life companion as a prerequisite for the formation and negotiation of a Muslim identity in a non-Muslim environment. We have seen how Cecilia met her husband-to-be quite soon after her conversion, through a Muslim way of "dating." In a similar way, Fatimah met the man who, a month later, would become her life-companion. Both before and after her official conversion to Islam at the religious center, she went through difficult periods of depression and loneliness. Becoming Muslim requires support from the outside, others to confirm the change, somebody with whom the person can structure, both psychologically and socially, a Muslim way of life. Like Mariam, Fatimah underscores that if she had not met her husband she would probably not have been able to "stay Muslim." Their gendered subjectivity is partly formed through the personal meaning of veiling and the relationship with their Muslim husbands.

Fatimah was lonely and anxious for help and a change. She describes herself as a "family person"; she needed someone close who understood her and with whom she could share and put into practice a new religious belief. Fatimah agreed to meet a man, an arrangement made by the man at the religious center that she had been in contact with previously. She knew how different this agreement was from the common ways of dating someone in the United States and—even more—this was meeting someone with the purpose of checking each other out as possible spouses. In a merry, laughing mode Fatimah recalls,

> *Fatimah:* He [the man at the spiritual center] asked, "Are you ready for change?" And I said, "Yes." And he said, "Do you want to get married again?" And I said, "Well, that is one thing that I really know, that I do want to be married, but I want it to be the right person." He said, "Well, I know someone that will be good for you.

Will you meet him?" And I said, "Okay, I'll meet him." "Don't you
have any questions?" he asked. And I'm trying to think what should
I ask. And he said, "What about where does he live? What does he
look like? Does he have any kids? Back to this world, Fatimah!"
"Okay, about him, well I want to know, is he a gentle person?"
Because my first husband wasn't what I would call a gentle person.
"Oh, yes, he's like a lamb." "Well, is he strong, because I don't want
any who's a weak person who is going to follow me." He said, "He
is strong as a bear." Okay, somebody who is like a bear and a lamb
[laughs]. I said, "Does he like children?" "Oh, yes, and he has
already two children and I know he wants more." And I said, "Is he
patient?" Because I was a new Muslim and I don't want somebody
coming making this hard for me.

Anna: Right.

Fatimah: "Well," he said, "he was married to a woman before for
twelve years and he is extremely patient." I said, "Is he a good com-
municator? I need someone that can talk." He said, "Well, he is
quiet, I don't know, you have to find that out by yourself." I said,
"Okay, I'll meet him." And then at around three that morning, I'm
lying in bed and I sat straight up, "what have I committed to?"
[laughs]. Because when I said that it wasn't just I'll meet him but I'll
meet him with the idea that maybe we will get married. It was a
strange sort of situation, and one which is very unheard of in
America. So I went back home and about two weeks later I got a
phone call, and I went to the phone and it was this man. All I knew
was his name. I didn't even know if he was black or white or Arab
or American. I didn't know anything about him . . . And the first
time I saw him, when he came to the door and came in I knew, literally, that within ten minutes of meeting him, that if he was going
to ask me to marry him, I'll marry him.

Anna: Yeah?

Fatimah: It was just in his eyes. And from his perspective, he had been
doing prayers, all the way from northern California, so I'm sure that
there was a lot of light in his eyes. He was wonderful.

When I later asked about the possible doubts one could have as an
American woman to an arrangement like this, Fatimah replied,

Well most of that [thinking] . . . disappeared. I was caught in this fairy
tale of perfection . . . You know there were all these things that were
happening that were so perfect that made me feel like I didn't have a
choice. You know, it was so meant to be that you can't even begin to
doubt—

Anna: —just follow.

Fatimah: Yes, just follow. Get into the boat and go.

In Fatimah's world, God had it perfectly planned. God had sent her a boat with a predetermined destination and there was nothing else to do for her than to climb into it and be led to whatever was meant to be. To let the boat pass her and float away empty, without her, was not even an option. Once again she felt as if she was given signs from above.

At the time of the conversion, and also when marrying, Fatimah said that she would never wear "the scarf," as she frequently refers to it. Just like all of the other women, she experienced it as one of the most difficult things to adjust to being a Muslim. Today she does cover, but as she puts it herself, her way of covering is somewhat of an "American adoption, kind of a compromise because it is not quite as covered and yet you are still covered." The scarf, tied in the back, covers her hair but not the chin and neck. This is what she is most comfortable with. Her initial resistance to wearing it was partly because it was more or less impossible to "blend into a crowd" and partly because people's judgment of the arrangement of her marriage.

> *Fatimah:* I felt very vulnerable, I was suddenly . . . I had only known him for forty days before we married so it was very quick. And I felt like I was afraid of people having a judgment about that.
> *Anna:* The arrangement?
> *Fatimah:* Yes, the arrangement of our marriage and the speed with which we got married. You know, I mean, it is a wonderful story how we met, but you don't go telling everybody the story. And even if you do they don't believe it.
> *Anna:* They don't believe it?
> *Fatimah:* Well, they believe you but they think you are crazy. They think you are crazy to rush into something. This is the age of trying it out and see after a couple of years if you . . . you know.
> *Anna:* How do you look upon it today? The arrangement?
> *Fatimah:* It was wonderful. To me it was such an exciting thing because we didn't kiss or hold hands. So it was exciting. It was such a different experience to have this anticipation there. You know that is not part of our culture anymore.

During Fatimah's reflection of the changes in her life, she draws on Muslim representations of the relationship between man and woman and how to dress and contrasts them to general "American"/"Western" representations and customs. The pleasant thrill of not having any deep physical contact before the marriage is compared to the Western way of getting to know each other, living together before any official commitment. Fatimah and her husband had both been married before and had children from previous marriages. One major agreement made before their marriage was "that we both felt that children were important and that I would stay home with the kids and be a mother, which was very important to me at that point." The salience of motherhood to Fatimah is supported by a religious discourse as well as a religious community.

Fatimah was not the only convert that elaborated on the novelty and attractiveness of the notion of being nice only for her husband and no other men. It seems as if, after having explored sexuality in different relationships, the idea that all sexual attention and attraction should happen and be channeled within the marriage is attractive and tempting in itself. It also brings with it a sense of safety and security, maybe missed in the previous lifestyle and relationships. The conversion, as the donning of the veil and the formation of a female Muslim identity, has in that regard resulted in a solution to a previously conflicting and unsettling situation, sometimes as a resolution of loneliness, fear, and yearned-for stability.

During the twelve years of being a Muslim, Fatimah perceives herself as becoming more conservative. She "strongly values family," she has home school for the two younger kids, she wears the veil and is critical but not judgmental, as she emphasizes, of the revealing clothing of American women in general; clothes she once used to wear herself. Fatimah has appropriated a Muslim discourse about femininity and female sexuality, but gives it a particular idiosyncratic meaning. Based on personal experience, she sees "conservative dresses as protection rather than oppression." She knows she has changed fundamentally in this regard, seeing the clothes she wore fifteen years ago as "dangerous." "I really see it as a way of exuding sexuality and tempting men." When newly married she started a new life in a new place with a new man by her side. She was happy but at the same time she felt very vulnerable because "everything had happened so quickly."

> And it was so new and so different from anybody else's story and how they had got together that I found myself getting up in the morning putting on the scarf in my own home and leaving it on until I went to bed at night. Feeling like it was truly some kind of cover for that vulnerability. It was somehow protective.

She experienced the scarf as a protection from the world's judgment of her different story, her new, divergent identity and life choice. Compared to her previous lifestyle and marriage, Fatimah now had a new role as wife and mother-at-home. She spent most of the time within the home, except for when picking up the kids from school. There was not just an inner transformation, her interaction with the outside world also altered. Before the marriage she had taken active part in public life, now she perceived it partly as something to protect herself from. Not in the sense that "there is that big evil lurking out there that is going to get me," but more that there is something *inside* herself that she has to protect herself from when in public spheres. Fatimah offers an interesting understanding of her relationship to the scarf, an understanding pointing to the veil as a personal symbol.

> *Fatimah:* I feel like it is . . . personally it feels like a protection . . . hmm, protection is not a good word . . . somehow wearing a scarf creates a space. And in that space I can be [silence] . . . safe. Maybe it is protection, I don't know. It is a barrier. It keeps unwanted energy away and I think that on another level it is intended to keep spiritual energy in rather than out.
>
> *Anna:* What do you mean by unwanted energy?
>
> *Fatimah:* [Silence] I don't really know. On a very mundane level it keeps male energy away from you. You don't find guys eyeing you which, you know, as females we kind of want that and don't want that. You want to be attractive at the same time as you don't want people that you don't want being attracted to you [laughs]. So in that respect it is both good and bad. It keeps all of it away and as a married woman that is better. So in that respect it keeps, not just other people's energy away from me, but it keeps *me* from being in that energy of trying to be attractive to other people.
>
> *Anna:* You don't have to deal with it.
>
> *Fatimah:* Not even dealing with it, it is more like . . . Many years ago, after being divorced from my first husband, I found myself wanting male energy, wanting to meet someone and in that process of wanting to be attractive . . . there is . . . I found in myself, I'm not saying for anybody else, but for myself I didn't always make the best judgments. And that I had a side of myself that liked to play with that very much, to receive that energy very much. In wearing a scarf it keeps me from stepping into that place which I consider as a very dangerous place in myself. Because if you are in a marriage and you find yourself, for example, frustrated with something that your mate has been doing the last week or month or you are very unhappy with one aspect of it you have a choice to either work within that marriage and stay contained in it and work through it as a family or, as in these times it is very easy to start looking outward which is how, I think, marriages fall apart.

Anna: You let yourself be distracted to something else.

Fatimah: Right. Or you find yourself happy, you don't go through it, you go around the issues. And somehow the scarf for me keeps me from even thinking that way. It keeps my energy focused on the fact that I am in a marriage and the person that I need to be most attracted to and for is here in my home. And if he and me are pleased with each other I don't need to be worried about if somebody else thinks that I'm attractive or not. And it keeps the energy much more focused at home, which I personally believe, and this is one of the areas where I think I have become more conservative. That this is a good thing. That marriage is a sacred place and we don't need to put that energy out. I know myself well enough to know that I have that, I could, I would. I can be very outgoing and putting my energy out there so when I say protection this is lot of what I'm talking about, protection from that place in myself.

Fatimah understands the scarf as something that separates her from the world and the temptation from others, from a sexual game between men and women, but also from a part of herself where she finds female sexual energy, a part that she recognizes well. Sexual attraction and temptation is contemplated upon through the concept of "energy" that can occupy a place inside herself. This echoes in part Mernissi's discussion about female sexuality as active. The scarf constitutes a safe space, an inner discipline that protects her not only from "male energy" but also from her own desires and need for attention and being attractive. She defines it as a "barrier" that keeps her away from a place within herself. Fatimah understands it partly in terms of sexual energy that should be kept from and outside herself and spiritual energy that should be kept inside. Most importantly, her own energy should be focused on her marriage and the person she is married to, who is the only one she should be attractive and attracted to. She points out,

> It is a reminder of Islam, it is a reminder of my practice and it contains for me my home, my modesty, and my privacy. And it is a constant reminder that I have made a choice to be Muslim and to be married. So for me it is a part of who I am now.

Two divergent images emerge through which she captures her self-transformation. By portraying the contrasting identities of the Western woman, exposed to an open sexual game in public space, versus the covered, modest Muslim woman mostly within the private space, Fatimah successfully uses "not-Me representations"

(Gregg 1998:144f.) in her self-understanding. The antirepresentation of the scantily dressed Western woman was herself before the conversion. The "previous" Fatimah was running in short shorts, wearing short skirts, receiving desired sexual attention from men, while today's Fatimah is covered and protects herself from unwanted male energy and from her own sexual energy. (See also the discussion in chapter 3.) In her earlier marriage she was working full-time keeping her children at daycare and in school, but today she understands her role to be, above all, that of a mother and wife-at-home running home school for the children. However, a year before our first interview Fatimah had started to help her husband in their business. This has involved her being "public" again, now having to respond to people's different reactions to her Muslim identity (see chapter 8).

In her account different cultural ideas of the organization of gender and the definition of freedom are expressed. But we should be careful not to interpret Muslim organization of gender based on Western terminology and definitions. As Ewing argues, gender is often spatially organized in Muslim societies, resting on a distinction between "inside" and "outside." While Western feminist discourse tends to associate "outside" with freedom, there is a dominant idea within Islamic discourse that situates power "inside," at the center (Ewing 1998:264).

Fatimah understands the veil as something that keeps the spiritual energy inside. An image of veiled power emerges; the experienced space of safety implies a sense of control and power. The power of being "invisible" but observing, the power of controlling sexual energy from both outside and inside. Understanding the veil as "protection" from "guys eyeing you" instead of as oppression, she thus reverses the power perspective (cf. Alloula 1986). By donning the veil she disciplines her own body and sexuality as well as strengthens her Muslim sense of self. Also, visualizing a Muslim belonging and identity brings with it another kind of visibility, her Muslim identity is under constant display and many of the women stressed the importance of acting as a good example. There is then a vulnerability of judgment and inspection from Muslims as well as non-Muslims.

Islam and Sufism permeate the thinking and doing of Fatimah, offering her life meaning and longed-for stability. Her self-understanding is organized around an *order/disorder* distinction comprising the contrapositions of spirituality/materiality, sacred/profane, private/public, female/male, morality/immorality, and natural/unnatural. The veil marks and maintains a necessary order between men and women. It is a demonstration of female modesty and of moral and sexual order, sustaining separation between the genders and their complementary roles. The order of the household should be a smaller impeccable reflection of the order of the society, without which chaos and confusion would arise.

> I think in Islam it is strongly recommended for women to be home with the children, it is not a law or a rule but it is promoted. There are so many blessings and *hadiths* saying that it is beneficial for everyone to have roles that are designated. That the man is the head of the household and I think that is another concept in the West that we are very opposed to. Here man and woman are said to be equal, which means the same, but I don't think that equal means the same. I think we are very different and that we have different roles, and in a perfect society, in a perfect home, and in a perfect household you still have someone that is head of the household making certain decisions.

Unlike Marianne, who has placed the Muslim ideal of a mother-at-home, giving her husband the last word, in dialogue with earlier ideas about gender equality, the commitment to women's rights, and her own desire to work and be engaged in the public world, Fatimah has internalized the Muslim representation of femininity and gender roles as a dominating, moral message. Consequently she also rejects her previous lifestyle, using it as a contrast to her role today as a Muslim wife within the "sacred place" of marriage. Freedom, to her, is reflected in her choice to be home with her children. In her previous marriage this kind of decision was neither encouraged nor appreciated for financial reasons. During the short periods of time when she did stay at home, she felt that she was not worth anything to her previous husband. On the contrary, he valued instead her financial contribution to the family. Today her strong identification with motherhood, embodied partly through the veil, is encouraged and confirmed. And consequently, religious messages on gender, such as ideas on complementary gender roles, sexual segregation, and female modesty, are reproduced.

In my discussion I have shown that veiling and modest behavior serve as means to meet particular personal quests while at the same time upholding cultural and religious norms. As a personal symbol, *hijab* embraces meaning on a social and cultural level in terms of widely shared Islamic ideas about gender complementarity, the recognition and control of female attractiveness and sexual desires in order to sustain social and moral order, as well as, on a psychological level, it addresses personal ideas on sexual energy and protection. In Fatimah's world, the veil is her modesty and privacy. It is her home. Inseparable from her self-understanding as a modest Muslim wife and mother, the veil is, in her own words, a part of herself.

To the converts, veiling and sexual segregation imply not only protecting themselves from sexual desires but also protection from being sexually objectified. As so manifested in the accounts of not only Fatimah and Cecilia but also of Hannah, embracing Islamic messages on gender relations and donning the veil become successful repudiations of previous lifestyles. Their arguments echo the ideas mentioned above about *fitna*, the chaos caused by female sexuality when it is not controlled, as when jogging in tank tops and short shorts. Being aware of what they understand as their power of sexuality, both Cecilia and Fatimah know well the impact they can have on men. By covering themselves they achieve the power to control and resist the male gaze simultaneously as they reproduce and enact certain Islamic ideas on gender.

Exploring New Femininities

Besides manifesting one's relationship with God, resolving personal concerns, and symbolizing affiliation, the veil also assigns the body a certain role. Veiling is a bodily practice and in the public sphere social criticism is formulated through the body. The Muslim dress code has been described as a "bodily reminder" of their faith, of their commitment to God, and of what is expected of them as Muslim women. The bodily experience of wearing the veil, and the strong sense of solidarity with other veiled women that this experience brings with it, reinforces their Muslim identity. It might even become an inseparable aspect of the appropriation of the Islamic doctrine itself (Brenner 1996). In a non-Muslim environment, it signals to the surrounding world the women's new identity, which does effect the way people interact with them.

The veil embodies an underlying dimension of the women's sense of transformed femaleness. As we have seen, veiling is intimately related to notions of self. For Fatimah, the scarf is perceived as an embodiment of motherhood. It reminds her of her Muslim identity and her role as wife and mother. There is a sense of personal integrity expressed here, or as Hannah put it, "a woman without the veil is like a woman without her purse." Many of the women expressed throughout their narrative that becoming Muslim had brought with it an increased feeling of self-esteem and respect. The veil or scarf as an inseparable part of the women's Muslim identity seems to enhance this feeling of pride and self-confidence, despite the negative labels attached to it.

Strong criticism is ventilated by the women against the public exuding of sexuality and what they see as a Western objectification of the female body. Some of them also pointed to their right to diverge from Western ideals of appearance, their right to gain weight, and not to be personally judged for their appearance as a woman. One of them told me that since her conversion and marriage she weighed twice as much as she did before. "But my husband should love me as much anyway." The women's accounts manifest social criticism against the public exploitation and objectification of the female body and representations of sexuality in the West. Veiling is a reaction against an opposite practice, which they consider just as degrading and suppressing as the general Western view of the Muslim veil. Here the veil becomes a bodily marker that encodes and communicates particular social messages.

The converts have refashioned themselves as female Muslims in different ways. Some of them are drawn to the salience given to motherhood and family ideals while some, who are all Swedish, echo a dominant public discourse on gender equality and criticize the traditional Muslim female roles while at the same time avoiding the perspective and arguments of Western secular feminism. In general, an external representation of femininity and gender equality seems to engage most of the Swedish women in quite strong emotional ways and in the interaction with a Muslim belief these are expressed and elaborated in various ways. Like Muslim feminists, they stress the egalitarian voice of Islam (cf. Ahmed 1992). By placing two different internalized representations into dialogue with each other they construct a new sense of femininity that emphasizes equality and their rights and in some cases the biological differences between the sexes. Criticizing the cultural ideals and demands on the Western woman and the objectification of the female body, simultaneously as donning the veil, the women manifest a critical commentary on cultural gender norms. The formation of a female Muslim identity, drawing from diverse public gender representations and the personal meaning given to the veil, suggests an examination of transformed and alternative femininities. In the social and cultural contexts of the women, the veil is understood in terms of sacredness, resistance, personal integrity, and power.

Part III

Encounters

Chapter Seven

Looping Effects of Meaning

The women are frequently objects of offensive and insolent verbal harassment and looks. The modest Islamic clothing seems to be the main reason for these insults and reactions. In encounters with strangers this is the only visible thing that communicates their Muslim identity. With the veil and long, loose dresses sometimes covering everything but their faces and hands, the converts are often placed within categories they do not necessarily identify themselves with, triggering a gap between self-identity and social expectations. Having their identity as a constant target for judgment, suspicion, harassment, and curiosity brings forth sensitivity in regard to their sense of self. Being viewed as "different" and having to relate to this difference is an unavoidable and essential part of the convert's identity.

In the previous chapters I discuss the cognitive formation of Muslim identity and how the conversion involves a personally recognized internal transformation as well as continuity of self. In this and the following chapter I focus on how the women relate their sense of self to the surrounding world, how they negotiate their Muslim identity externally, in different social interactions. By attending to the women's experiences of the manifold encounters and interpersonal relationships in their everyday lives, we can analyze the different expressions and shifts of their self-understandings in relation to others. Becoming a Muslim involves not only an inner sense of change but also a socially recognized one, as well as a change of social role and environment. The encounters are of interest not only in terms of the shifting sense of self in relation to others, but also because they reflect the social contexts in which the converts live and act, and the different attitudes and strategies employed in a sometimes hostile environment.

The interactional "others" that the women talked about are parents, family, employers, other Muslim women, female Muslim converts, neighbors, strangers, an "imagined nation," and oneself.

In these interpersonal encounters, diverse cultural labels and representations and their meaning come into play. It is in these social interactions that the women try to communicate and negotiate the personal meaning of being Muslim as well as the labels assigned to them. Different categories are employed, rejected, accepted, and modified by the individuals involved. The encounters spark self-reflection and trigger several possible self-understandings and strategies inciting feelings of belonging, not-belonging, national identity, and a religious self. In these face-to-face interactions, different meanings of "Swedishness," "Americanness," "Muslimness," and "femaleness" enter into dialogue with each other, sometimes bringing out conflicts and struggle, sometimes harmonization and agreement.

Targeted Selfhood and Transitional Meaning

During our conversations, Marianne elaborated extensively on her experiences of different interpersonal relationships and how she experiences herself within them. What emerged from her narratives about her relationship to the surrounding world was a targeted selfhood. She expressed a psychological unease in regard to the misunderstandings and crude labeling of her as a Muslim woman. During the interviews she often looked upon herself from the perspective of others, expressing a heightened awareness of various face-to-face confrontations and what they did to her sense of self. Similar to the other women, Marianne described the first times she visualized her Muslim identity in public as particularly difficult and uneasy. It takes a long time to get used to wearing the veil in a comfortable and relaxed way in a non-Muslim environment. Below Marianne gives an account of moments of self-reflection, or, "encounters" with herself.

> I remember I was standing in front of the mirror trying the towel [like a veil] after I had showered. And I looked at myself in the mirror and thought, "What the hell do I look like! Is it possible to look that silly?" I looked at myself and thought that I really didn't look quite right in the head . . . For many days I sat at home and finally I went outside [with the veil on]. I was very nervous and I was looking at everybody.
> *Anna:* I can imagine it must have been difficult.
> *Marianne:* Yes, it was very hard if you met anybody you knew from before.

Anna: I was thinking about encounters with old friends.

Marianne: Yes, it is very hard. I withdrew for quite some time. I had a friend and I believe she felt that I was withdrawing. And some months later I got pregnant and I stayed home. Then I got more Muslim friends. There were many [of the old acquaintances] who socially disengaged. Then I got it into my head that they didn't recognize me so I didn't bother to greet them, because I imagined that they didn't know who I was. But they knew very well who I was. These were old classmates with whom I had rather passing contacts. [Her children come in and we take a short pause.] But then . . . it was rather difficult to be Muslim. You weren't either really sure in the beginning what you believed in. You didn't really have any answers; it was more that you didn't really know why you had made this decision.

Anna: You don't know for sure?

Marianne: No, sometimes you think, "is it a dream?" "what have I done now?" And sometimes it hits me when passing a window and I see my reflection. Then I can think, "is that *me* who looks like that? That looks that strange? Why have I done this?" It still strikes me how weird I look with the veil. I look strange [laughs a little]. It is another personality that I'm startled by. "Is that me?" I don't think of myself as a Muslim all the time. I see, ugh, how obvious it is.

Marianne reflects here upon her transformation. Confronting her own reflection, Marianne perceives herself from the other's point of view, or maybe rather, from the perspective of Marianne before the conversion. She looks upon the apparent change in her life and her visible Muslim identity. She verbalizes an inner dialogue spurred by the experienced strangeness of self. "Is it a dream?" "Is that me?" The reflection in the mirror and in the window, of who she has become, of who she is today, is looking back on the narrating "I," a reflecting self, who embodies many different self-presentations, diverse experiences, memories, and feelings of changes as well as continuity in life.[1]

In many everyday situations Marianne's Muslim identity is not necessarily a matter of focus or special attention to her. In that sense, she can almost forget about it, or in Mariam's words, it has become "natural." As Marianne asserts, "I don't think of myself as a Muslim all the time." Disparate experiences and thoughts have been integrated into a conceivable coherence. But now and then, as in the narrated situation above, she is reminded how "strange" and different looking she might be for some people in her surroundings. Appropriating an expected attitude of other people and projecting it on her own self-reflection, the clash between her self-image and the social and public categorization of her becomes evident. Marianne's feelings of who she is do not correspond to the connotations and images her appearance brings forth, triggering what Linger (2001:14) has called "reflective consciousness." In his study of Japanese Brazilians in Japan he demonstrates how their sense of self becomes an object for reflection due to the experience of dislocation. Similarly, identifying with targeted identity categories brings about moments of heightened self-reflection on and sensitivity to who one is in relation to others.

Anthony Cohen (1994) has problematized the disparity between the person's subjectivity and the socially and collectively imposed categories or identities, distinguishing between selfhood and person-hood. The experienced "tension between selfhood (the substance of 'me' of which I am aware) and personhood, the definition of me as a social entity which society imposes" (p. 56f.) is a part of the social experience of the converts. Society imposes particular social cate-gories upon the convert. Socially, she is perceived differently. She has most likely become involved in a Muslim community and organiza-tion, has new ties of solidarity, and has perhaps even become a statistic somewhere as a "Muslim."

Social interactions stimulate reflexivity and engagement with cultural and personal meaning. The continuous meaning-making taking place in social interactions reflects transitional meaning, namely, meaning is not inert but altering and passing, eliciting new interpretations and reflections. The women are not merely shaped or determined by the interactions; they are compelled to engage people and assigned categories and labels (cf. Linger 2001:306). The out-come and meaning of these face-to-face encounters is not predictable per se or simply explained by the "positions" to which the women are ascribed. The women are sometimes more or less forced to inter-act with social expectations and public representations of how a "Muslim," as well as a "Swede" or an "American," should think and act. They may sometimes act in accordance with expectations or resist, question, and choose to work back upon the imposed classifi-cations in a provocative, sometimes capricious, way with the aim of reversing the power relation of the situation.

In a similar vein, the philosopher Ian Hacking describes this capacity of meaning-making and the "looping effects" it engenders. In *The Social Construction of What?*, Hacking (1999) has an excel-lent discussion about the differences between the constructions found in natural and social sciences. The objects of study within social sciences are most often of interactive kinds. By this he means that the idea that is described as constructed, let's say the construction of the idea of the "Muslim woman," interacts with the individuals who fall under that idea. Hacking's criticism of "construction talk" is that "it suggests a one-way street." He continues, "By introducing the idea of an interactive kind, I want to make plain that we have a two-way street, or rather a labyrinth of interlocking alleys" (1999:116). With an emphasis on the aspect of process and interaction, he discusses how classifications can influence what is classified as well as how people react to and sometimes change the classifications of themselves, and as a result how the classification may be modified or replaced. "People classified in a certain way tend to conform to or grow into the ways that they are described; but they also evolve in their own ways, so that the classifications and descriptions have to be constantly revised" (1995:21). He calls this the *looping effect* of human kinds.

Hacking argues that the target in social sciences is always on the move. Social sciences consequently deal with "moving targets," (1999) namely, that self-aware, reflective people who are the objects of the inquiry may understand how they are classified and rethink themselves accordingly or change in another direction in response to being classified. Even if Hacking is specifically interested in social and historical conditions of phenomena such as multiple personality, child abuse, and homosexuality, I have found his discussion quite constructive in my own analysis of shifting and transitional meaning-making. In his criticism of social constructionism he questions whether these phenomena existed before they were conceptualized as such, noting the "looping effects" by which people can react to social science descriptions and categorization by, for example, acting out and upon such descriptions. I use looping effects in a broad sense referring to the idea that through a reflective self, meanings of categories change when they are applied to real people in real situations.

While I find quite fallacious the widespread assumption that social and public ideas are automatically and indiscriminately reproduced in social life, this study points to an opposite notion. The encounters and confrontations in the women's everyday lives highlight the dynamics of classifications, the looping effect of meaning, namely, how imposed categories and their assigned meanings trigger self-reflection and are rearranged and redefined by people, as well as how feedback can direct itself in various and unpredictable ways. Within this dynamic and incessant process of transitional meaning, in which meaning is bounced back and forth and therefore transformed, the women interact with, challenge, and explore the

t interpretations and understandings, such as the socially
cted *ideas* of "the Muslim woman" that come into play.
Similar to my analysis of personal models in previous chapters I will
show how, as soon as a public label or category is tried out and
engaged, a *personal commentary* is spurred. Inspired by the idea of
looping effects of human kinds I will demonstrate, by looking closer
at Marianne, that when a representation or an identity category is
appropriated by a person in order to negotiate a sense of self, its
meaning is frequently modified.

Marianne's Shifting Use of Labels and Identity Categories

Below is a longer, somewhat rearranged excerpt from the interview
with Marianne pointing to the different, possible self-understandings
that emerge in various encounters. The reader can follow how
Marianne relates herself to, objectifies, and engages different categories such as "Swedish," "Muslim," and "immigrant" in rather
shifting ways, depending on the context. The long interview excerpt
allows the reader to get a feeling of the interview situation and make
her/his own interpretation of the dialogue.

> *Marianne:* I think it is hard to encounter people who have pictured me
> as Swedish. Once I took a class and I had had contact with the
> teacher over the phone. Then when she saw me she became so per-
> plexed. She said nothing. She stared and couldn't . . . Then later she
> called me and apologized. She somehow couldn't deal with the fact
> that I was a Muslim. It took her some time to get over it because I
> sounded like something completely different over the phone. I think
> that is difficult. I always send out a warning to people that "I'm
> Muslim too" . . . I think somehow that I want to be considerate to
> the other so she or he doesn't get that awful feeling of "ugh, I can't
> handle this that she is Muslim."
>
> *Anna:* It is difficult for both of you.
>
> *Marianne:* Yes, it is difficult for both of us because people become so
> disconcerted. And I usually show consideration by telling them in
> advance. It got so silly when we were about to buy the house. I told
> the real estate agent in a kind of apologetic way that "we are
> Muslims." "Yes, yes, as long as you have money," he said [both
> laugh]. It became so silly and forced. People can maybe also become
> irritated by me saying it. Because they think that they can deal with
> that. They maybe don't think it matters. For example, the neighbor

here asked me where I'm from. "Well, from Sweden," I said half
annoyed. "Yes, but where in Sweden?" he wondered . . . There are
misunderstandings all the time and you presuppose a whole lot. And
you can't really categorize me, nowhere, neither here nor in the
Arab world. There will be more strange identities, mixed identities.
My children will be more Swedish than I . . . I don't know.

Anna: You say more Swedish and you say that she thought I was
Swedish . . . and you are Swedish, right?

Marianne: Yeah, I am, it is true. But I identify with immigrants all the
time. There was another convert who said, "so I sat there waiting,
another immigrant and me." That is, that she was an immigrant too.

Anna: You use the category "immigrant," which is quite broad. You—

Marianne: I sometimes feel like I'm both. But I think that I belong to
the "immigrant side" more than the "Swedish side," most often.

Anna: Has it always been like that [since you converted]?

Marianne: I don't know. I guess there has been a change. You can
probably accept the Swedish things more now than you could in the
beginning as a Muslim. Then you were only Muslim and Swedish
nationality, what the devil is that? It is nothing and you repudiate
everything to do with nationality. And that is also something you
are supposed to do in Islam. That nationality shouldn't matter, that
is the ideal. But in one way or another you can't escape, you can't
reject who you are, you can't get rid of what you are, what you have
thought and what you feel.

Anna: How do you relate to the Swedish aspects?

Marianne: Well, both, right? I think it is good that I have the background I do. I think as a Swede you have learned to see a lot that I
think Muslims often are totally blind to, for example, this regarding
gender issues and those structures.

Anna: And that you have been able to combine with Islam. And this
about a political standpoint seems to be something many can—

Marianne: Yes, do you have a socialist background, you can see it
relate to the thought of solidarity, you have the alms for the poor[2]
and you have the idea of equality in Islam. You can take what you
want, right.

[. . .]

Anna: I was thinking of what you said earlier . . . that you identify
with immigrants.

Marianne: But it also depends on the context. Among Muslims, I'm
100 percent Swedish to be sure.

Anna: It depends on who you are meeting. You still see yourself as
participating in Swedish society like elections and such?

Marianne: My husband for example thinks it is idiotic to vote. He
thinks there is no point in voting. It is just silly, in his view: you
shouldn't participate in the society. I have been voting over the
years . . . But I also know converts who don't. But . . . yeah, so it is

different. What was most emotional for me was to go back to the countryside, to my childhood place. To get there and be so damn different as a Muslim. It is very difficult to be a Muslim there. They "ha, ha what do you have on your head?" and "we have, you know, seen you since you were small so take that shit off." "Are you going to wear nightshirt the whole day!" It was hard.

Anna: How do you tackle that?

Marianne: The first year I couldn't cope with going there. I was so very saddened by what people said. But I also know it is kind of banter. But I get most sad about always being an outsider there. That I can't come back and be the one I have been. When you are there, you go and eat at people's homes. Well, and then you can't eat anything. It is always like that. It is always pork, or some kind of other meat, or some kind of margarine that you don't really want to eat. And then there is alcohol there. I'm culturally entirely outside; I can take part some and I can show consideration to a certain limit. I mean, as a Muslim I don't drink liquor and actually I shouldn't be in contexts where people drink either. Those are the small things. I can't go in and dictate but I do feel that here I have to draw a limit. Then I feel that I have to choose to be an outsider. I always have to draw a limit or that they have to think twice. So then I think that there is a border and that is maybe the way it should be. Because I can't go inside and pretend like I'm like them. On the contrary, it is all the time a question about eating . . . and clothing and nakedness.

Anna: Do you feel it is easier now after some years?

Marianne: Yes, now I can . . . partly, I'm older myself. I'm not as sensitive about all of those things. But it is still difficult. It is much more difficult there. You know I lived in the city when I converted and I didn't have any permanent friendship with anyone. There was no problem. But there [at her childhood place] it is difficult.

Anna: I was thinking if you get used to what strategies you could develop?

Marianne: Yes, you do. You do, sure.

Anna: And they have got to know you or have they?

Marianne: It has become quite OK. My children went once to the swimming school there and it was the last day and the newspaper was there and took pictures. And then it almost felt like they were proud of us. Because here they indeed have some strange ones too. I know my aunt sometimes took my children to the swimming school and she sat there by the pool saying, "Yes, I'm counting darkies," counting my children so no one had drowned [laughs]. I accept her saying that. I didn't mind, I thought it was pretty funny. We live by a cove and they renamed it the "Persian Gulf" [laughs]. So I think that we do meet and connect, right . . . it works but it never feels like I'm really involved . . . My aunt tells me that she thinks it is difficult because we never get really close to each other anymore,

like we did before. There is something in-between . . . It has to do
with solidarity. That is what is hard. I really love this place and then
at the same time walk around there and be so damn odd, weird, and
strange. It becomes pathetic to be a Muslim there. It is different in a
town. I can't talk to them like I can talk to you. It doesn't work.
I don't mean that they are more stupid but they are in their world.
You can't discuss the fact that you actually can convert.

Anna: Is it because of your faith or is it your clothes or is it because
they expect and know that you should be different?

Marianne: I believe it is all of that, that I'm supposed to be different.
My faith is not that important, well, they do take exception to this,
they say, "So, then we all are going to hell because we don't believe
the way you do?" They get very offended by that . . . even if I never
have expressed that thought. And in one way, that is what it is,
right? I can't say that Islam is as good as anything else. Because I do
think Islam is better than all the others. They say, "You think you
are better than us." And in one way I do believe that, because
I wouldn't have been able to be a Muslim if it was indifferent to me.
Because I do believe it is good to be Muslim, I do believe it is the
best. I would lie if I said that it is as good to be something else . . .
I also believe I have hurt them. Like it is not good enough what they
have or what I was brought up with. And I have disappointed them
by not being a certain way.

[. . .]

Marianne: It is about daring. To dare being oneself and to meet each
other. If both are insecure, both will withdraw. It is difficult for me
because I'm a rather shy person.

Anna: Really? [I show my surprise, that it doesn't really fit my impression of her.]

Marianne: I'm afraid to make a fool of myself. I realized that, that is
what I am. So I shouldn't have become Muslim because then you
really make a fool of yourself [laughs]. I'm afraid to fail. I don't like
when people look at me either.

Anna: But they do.

Marianne: Yes, that's for sure. That is one of the most difficult things
being a Muslim, that you are always someone people look at. That
is why I sometimes think that I should live in a Muslim country so I
could be normal for once. And not be questioned all the time. And
exposed to looks. I feel like people think that I cost money [relating
to common ideas about immigrants costing a lot of money for the
taxpayers].

Anna: There are a lot of things happening here. You apply what you
think they think and the opposite.

Marianne: Yes.

Anna: And it is inevitable. Does it happen that people approach you
to talk?

Marianne: No, very seldom. But they have a certain way to look at you. They may not mean anything bad with their looks but . . . That is something you as Muslim are more sensitive to in the beginning. Then I looked at people's faces and their reaction. I took it to my heart all the time.

[. . .]

Marianne: [I have asked Marianne about the changes in life that the conversion has brought with it.] I have got a fantastic life as a Muslim. I have got engaged in good causes like helping Muslim women and so on.

Anna: It is not "only" a religious identity, is it?

Marianne: No, it isn't, I have also become an activist among other things. I feel like I have met a whole lot of fantastic people who I maybe wouldn't have met otherwise . . . I have a lot of social contacts which I wouldn't have got otherwise . . . As a convert you meet other converts and it is only with *them* that I really feel at home. I don't feel at home with Arabic women directly. If we are talking about identity . . . I'm almost exclusively together with converts. I also meet these immigrants who have become very Swedish. It really has a big impact on who you are and how you see things. I feel that very Arabic women are so damn boring.

Anna: By very Arabic, what do you mean, religious or—

Marianne: No, no, I mean those who are very much meshed in their culture, who haven't let much of their culture go and who haven't taken in much of what is Swedish. I also know an Arabic woman who has become, not Swedish, but she thinks in a very Swedish way. She is engaged in society, her children attend all kinds of activities and she drives them to these. She works . . . with them I feel totally at home. Then I feel Muslim but still in a rather Swedish way.

Anna: And with the very Arabic women you feel very Swedish.

Marianne: Yes, I think they have such boring interests. I feel like an outsider. I don't laugh at what they laugh at. Those who have become rather Swedish, if I express it like that, they also understand my jokes. It is about what one understands and what one refers to.

[. . .]

Anna: What about your encounters with your husband's family. How was it to go to their home country as a convert? How were you received?

Marianne: They treat you very well and they think it is fantastic that you have become a Muslim. As a Western Muslim you are praised to the skies. There are many converts, men that is, who have high positions in different Muslim organizations. Also women are rather active globally. They view you as something very amazing. They still have this inferior attitude toward the West and that they don't have anything to contribute. It is kind of implanted in their mind that they are worse, also because of the colonization . . . but then on the

other hand it all amounts to the idea that I should become like them. That is what I react against. I'm never one of them either. I'm an outsider there too, all the time. I can't speak the language and I don't want to speak their language. I have had this feeling that I don't want to. I prefer to be quiet in these situations. I don't know why. It must be something wrong with me. But I haven't had the energy to become engaged. I have studied Arabic but with their dialect it is so different. It is a kind of protection, a strategy that I have, that I don't want to learn the language because I don't want to be a part . . . because I don't like their culture. Maybe it is cowardly to do so but I'm not interested. I'm more interested to be in my own world when I'm there or when they are here. I don't want to talk so much. We can talk in the kitchen about everyday things, that I know in their language, but I don't want to get into deeper issues . . . Maybe it reflects an uncertainty on my part. I object to their idea that I should be altered.
[. . .]
I don't feel like I belong there. I don't belong anywhere but among converts, among Swedish converts. That's the way it is. I can never feel totally at home with them [her husband's relatives]. But with converts I feel like I can be 100 percent myself . . . To be able to totally be yourself. I don't think it is particularly Islamic what they are doing [her husband's relatives]. I have converted to Islam and not to their strange culture.

Marianne brings up a series of different encounters, face-to-face interactions with people from her childhood, her Muslim husband's relatives, Muslims in Sweden, Arabic immigrant women, neighbors, teachers, and other converts. What is really going on here? By employing certain representations and categories, she is trying to identify and convey the complexities of who she is, feelings of belonging and not belonging, of connection and separation, solidarity, loyalties, and betrayals. Reading the long interview transcript the reader can, I believe, get a good hold of the dynamic looping effects of human kinds.

Different interpersonal exchanges and relations generate obviously different self-understandings and sentiments of belonging. Her feeling of being Muslim, namely the meaning she assigns to her Muslim identity, is accepted and understood to varying degrees depending on who she is interacting with. For example, how does she convey the feeling of being different among Muslims and Swedes alike? Among Muslims she feels "100 percent Swedish" and with Swedes, like her relatives and friends at her childhood place, she feels "so damn different as a Muslim." Marianne oscillates between feeling "very Swedish" with Arabic women, very odd and weird with her relatives, feeling solidarity with "immigrants" in general, "Muslim in a rather Swedish way" with certain immigrants, and "100 percent myself" with other converts. It becomes apparent that to be Marianne, as with most of us, implies shifting self-representations, and that "Swedish" and "Muslim" have heterogeneous and shifting meaning depending on the context and people involved. Thus, categories do not have any definite or complete meaning; what "Swedish" means changes depending on the interactional others and the other categories involved.

By trying out different well-established national and religious identity categories and various expressions such as "100 percent myself," Marianne seeks to pinpoint and communicate her sense of self in relation to others. Frequently, similar shifting of self-representations are understood as resulting in fragmented selves, reflecting a theoretical thesis about the self as positioned in various sites by discourses. However, I do not think that the shifting from "Muslim," to "Swedish," to "immigrant," etc. can be taken literally as embodying multiple selves, a standpoint that often reflects the assumption that it is the multiple discourses that cause ephemeral selves. Asserting a person's sense of coherence and continuity, as I do in this study, does not mean supporting a theoretical idea about a monolithic, static, and bounded self, but neither do I, nor Marianne for that matter, understand her as a fragmented self.

There is another way to think through this problem, which problematizes prevalent metaphors such as the "border crossing individual" and the "fragmented self." What is it in particular that makes Marianne feel so "Swedish" in relation to Arab women and "at home" with immigrant women who have become "rather Swedish"? As I have stressed earlier in the book, I do not think that we should take for granted that the use of the concept "Swedish" necessarily draws on national belonging, or that it has the meaning of national identification (Linger 2001:309, 2003).* We have seen that Marianne's self-making is mainly organized around a salient personal model of gender equality and women's rights. These are understandings that engage her in diverse activities in her everyday life and through which she has appropriated Islam. When she is talking about "Swedish" values significant to her and when and how she feels "Swedish," these are the understandings Marianne frequently refers to. Marianne feels at home together with women who, like herself, take on an active role in society and appreciate gender equality, who work, and take their children to different activities. With them she feels "Muslim but still in a rather Swedish way." In comparison to her relationship with the group of Arabic women and her husband's relatives as well as her own, the rather "Swedish-like" Muslim immigrant women and other converts allow her to feel both "Swedish" and "Muslim" simultaneously, accepting both dimensions of her identity. More accurately then, I believe the different categories being used reflect to what degree she feels like she has been able to mediate her *personal meaning* of being Muslim. The oscillation reflects thus the looping effect of meaning triggered by the very ability of a self-reflective self, the ability to obtain a sense of coherence and continuity even while employing different identity categories, rather than a fragmented self.

The shifting of identities and the negotiation of difference has been of special interest for social scientists lately. In studies on transcultural identity the individuals are frequently addressed as border-crossing persons. The metaphor of border, a manifestation of the discourse of nation-states, can however, as Ewing suggests, limit us into one single way of conceptualizing difference. It is likely that it "forces us into a single discourse that does not adequately represent the processes by which individuals and communities think about and negotiate difference, ironically creating a modernist sameness in the midst of a celebration of postmodern border crossing and fluidity in the 'borderland' " (Ewing 1998:263). I agree with her, and moreover the idealization of the postmodern border-crossing individual can even—maybe implicitly, but efficiently—reinforce and confirm the borders themselves. The women do not always understand the differences they embrace as being so clear-cut or as "transgressing boundaries." It also raises crucial questions such as: borders? And *whose* borders?

The term "border crossing" itself presumes that there is an explicit set of borders to cross in the first place. This in turn implies that these borders actually encompass certain contents, certain ideas that differ from ideas in another encircled entity. In that sense, the talk about border crossing risks manifesting ideas about cultures as separate entities, for example, that "Swedish" means something particular in contrast to "Muslim." Similarly, the talk about fragmented selves often implies that there are particular borders to cross when shifting from one self to another, and that these selves, and the cultural categories they embrace, are irreconcilable. Just like cultures, "borders" are blurred and shifting. When Marianne uses the label "Swedish" she sometimes refers to ideas close to the self such as gender equality, while at other times "Swedish" signifies ideas she is critical of such as the public exploitation and manipulation of the female body.

As I demonstrate here, the coexistence of distinct and sometimes seemingly irreconcilable attitudes and practices is therefore not always easily illustrated in terms of crossing borders. For example, Marianne expresses her faith in Allah *and* her interest in postmodern thinking that there may not be any truth out there, Mariam and Fatimah integrate the Catholic teaching from their childhood with their Muslim belief, and Ayşe combines a "Swedish" socialist message with a Muslim belief in the good in people and solidarity with the poor. I believe it is more useful to understand the organization and negotiation of differences through the different discourses and categories people engage in and draw from and the personal meaning assigned to them. Differences are, as I have shown, often reconciled through a cognitive framework as well as put into practice within a particular personal life. They often make sense to the person and persons involved but may seem paradoxical for an outsider, who refers them to public discourses rather than the possible ways they may have been learned and integrated within an intrapersonal realm.

Let me return to Marianne's experiences. In the encounters Marianne's Muslim identity is in the spotlight, not only for others but also for herself. If her identity is questioned or she feels "exposed to looks," her first response is to wonder whether it is because of her Muslim identity. Like walking around on a minefield, she is cautious and highly sensitive not to trigger any explosions or provocations. In that sense, I believe, Marianne herself can make the encounters problematic and unsettling. Her face-to-face interaction with the neighbor and the real estate agent indicates this. The convert's Muslim identity does not necessarily have to be conceived as negative for the other person, but her own anticipations, interpretations, and of course her own experience that this is often the case can make some situations charged.

The transformation of becoming Muslim has implied a set of limits that she has to draw in relation to people who have known her for a long time, even with relatives she has been very close to. Things have changed. She has lost the connection with the place where she grew up and its people. Marianne loves a place in which she now feels like an outsider. She has changed and she cannot be nor pretend to be the one she has been. Her Muslim identity, her different faith, way of life, new appearance, and new loyalties are things unknown and strange to her family. "Their" Marianne has become someone else. That to which the convert converts is for them an unfamiliar

belief, and foreign habits have put an invisible but ubiquitous wall
between her and them. "It becomes pathetic to be a Muslim there."
The clash between who she was and who she is, between childhood
memories of connection and the present fractured relationships,
between feelings of belonging and of being an outsider, brings with it
a painful experience. Her relatives referring to the veil say, "We have,
you know, seen you since you were small so take that shit off." The
feeling of being "so damn odd, weird, and strange" to them brings
forth a sadness of loss. In this encounter her Muslim identity
"forces" her to choose to be an "outsider." In implementing the conversion and her new self-understanding as Muslim, it seems impossible to maintain the same kind of relationship with people as before,
relationships that were on their terms.

These are emotional encounters that took Marianne several years
to dare facing. She believes she has hurt them by choosing her own
way and, as a result, rejecting their way of living and also who she
was brought up to be. Different emotions are set in motion particularly with family and friends. There is a fear of disappointment and
a recognition of and identification with the other's distress due to her
own decision to become "strange." Internally, she can negotiate and
reconcile a sense of continuity and change, old and new ideas, but
externally, in face-to-face interaction such as those with her family
the change of self and life appears manifest and unavoidable. "I can't
come back and be the one I have been." She does not feel at home in
her old home.

In her Muslim husband's home country and together with his family and relatives Marianne likewise feels like "an outsider." "I'm
never one of them either." She strongly rejects what she feels are their
wishes and attempts to "alter her." She does not want to "become
like them"; she does not want to be a part of their culture. Some of
the crucial factors of this rejection concern rules and regulations
regarding the relationship between men and women. For example,
they do not want her to wear the veil inside the home but rather
show her appearance as long as there are only relatives present. She
nevertheless prefers to keep the veil on and also to stay in the room
even if a man from outside the family enters, instead of running out
to another room as the other women do. Marianne is not used to this
more restrictive separation between men and women and the limita-
tions regarding activities and spatial movement. She resists her
father-in-law's ideas about how she should look and act; she resists
his requests to have her dress up in "pearl clothes," like the "princess
clothes one buys in the toyshops. Frills and pearls." They want her to

shave her legs and show less hair on her body. She herself could not care less about this. Their views and the meaning they give to the veil counteract her own view and arguments for wearing modest dresses. As we see in the previous chapter, it is thus not only the Swedish cultural ideals of femininity that Marianne resists; she challenges socially constructed ideas of the Muslim woman both in Sweden and in her husband's home country. She feels as if they interfere and enter her personal sphere far too much. At the same time, she gains a lot of respect for being a Western woman who has converted to Islam, but she keeps a reserved attitude in her interactions with them. She consciously chooses not to learn their language and she prefers to stay, as she says, "in my own world when I'm there or they are here."

In many situations, Marianne identifies with "immigrants." Because of the veil, people also often categorize her as such or have difficulties categorizing her at all, "neither here nor in the Arab world." But it also has to do with her feelings of loyalty and of social and political affiliation and solidarity. She can identify with them as a minority group, as being "different," not looking or acting as a "Swede" is supposed to look and act. The use of the category of "immigrant" also conveys her feelings of social estrangement, not being able to convey her sense of self to the surrounding world.

Marianne does not really feel at home anywhere, neither in her childhood place nor in her husband's home country. Throughout her reflective account we realize that it is only with one particular group of people that Marianne feels at home.

What's strange is that we are totally flipped out when we converts meet. We laugh tremendously and are amused by ourselves. With the ones I know, we have some rather coarse banter. And that we can only have inwards, within the group; because seen outwardly we are enormously pious. But together we can relax. If I had been with the Arabic women like I'm with them, they would have found me awfully immoral and doubted me being a believing Muslim. We [the converts] know who we are, what background and positions we have. We can let out so many other sides that we hold back in other contexts, since it seems so strange if you are totally . . . I mean there are many who believe that I can't joke because they perceive me, as Muslim, as so very serious. Maybe I live up to that image to give the impression of being so very serious. It is only with converts that I can relax.

Anna: Why do you think that is?

Marianne: I don't need to be questioned. I can imagine if I was totally flipped-out with Swedes they may think, or maybe I think they think so, that I'm not a real Muslim if I'm like very ironic, the way I can be sometimes. Maybe I think that they think that I'm not really a believer if I act that way, being ironic about my own religion. When the converts meet we are very ironic at our own expense, both about religion and Muslims in general. That is what we joke about since it is so charged; that we are supposed to be so goody-goody, that Islam is so nice and then we see how the reality is . . . we hate those damn Arabic old men. We can sit and talk shit and that we would never do in front of Swedes or other Arabs.

Only with other Swedish converts is Marianne able to feel comfortable, only with them can she be herself. But what does it mean to her to feel, as she chooses to express it, "100 percent myself"? We know that Marianne understands herself as an independent, intelligent, and self-reflective woman, committed to feminist issues and engaged in several women's organizations. The circle of converts, close friends of hers, knows her as such too. Together with them her identity is not questioned or scrutinized. With them her sense of self, and her personal meaning of being Muslim, is easily conveyed and most likely, understood. "We know who we are, what background and positions we have." They confirm her own self-understanding. There seem to be a whole lot of charged and tense feelings being released when they meet. The gatherings have a seemingly therapeutic function; they provide social opportunities for her to air distress and interact in a way she would not be able to do with either "Swedes" or "Arabs." With "Swedes," Marianne asserts, she would not be able to make jokes and complain about those "damn Arabs." If she did, it would feel like she "served them arguments," such as "if it is that difficult to be a Muslim or if you do not agree with Arabs sometimes, why not just leave it all and have it like it was before?" There are obviously difficulties in criticizing a system to which you have converted. It becomes important to think about who you ventilate your problems and worries to. Within the small group of converts they recognize each other's anxieties and identities of being both "Swedish" and "Muslim" and the reactions it provokes, which create a powerful, emotional connection, loyalty, and understanding of solidarity. In this context there is no immediate and critical tension between her own self-understanding and the social image and expectations of her.

In her account, Marianne marks her degree of identification and solidarity with different groups by using the concept of feeling "at home." As Rapport and Dawson (1998:10) suggest, "[o]ne is at home when one inhabits a cognitive environment in which one can undertake the routines of daily life and through which one finds one's identity best mediated." Marianne's "home," her sense of belonging, is not where she grew up, but where she best knows herself (ibid.:9). "I don't belong anywhere but among Swedish converts." It does not imply the kind of belonging frequently conveyed with the metaphor of "roots" (cf. Malkki 1992). A particular physical or geographical place is not required, but rather an emotional state that is, in her case, acquired within a certain set of interpersonal relationships—with her convert friends. With them, her self as Marianne and the personal meaning assigned to her conversion and being Muslim are best mediated.[3] In connection to the transitional aspect of the narrative discussed in chapter 3, I also believe that Marianne perceives the interview encounters as fine opportunities to reflect upon these shifting self-understandings and the interactions between her and I as pretty satisfactory in terms of being able to mediate her sense of self and its complexities.

Reconsidering National Belonging

Becoming Muslim and the public meaning this identity tends to have for others have more or less forced the women to reflect on how their conversion and Muslim identity relates to national belonging and categories. The conversion stirs thoughts about national identity, a belonging that in many cases was previously taken for granted. Visualizing a Muslim identity, their national identity is frequently questioned by anonymous others. Becoming Muslim inevitably changes the experience of being Swedish or American, pointing to further examples of the dynamic process of looping effects of meaning.

Since the modest Muslim dress code signals a perceived cultural difference, people tend to place them in ethnic and cultural categories other than "Swedish." "We are looked upon as traitors to Swedishness," Ayşe tells me during our first interview. Similarly, Lisa thinks people see her as a "traitor to her country." She believes that people perceive her as an "immigrant woman." Once, a young man shouted to Ayşe, "Go home you fucking Arab hag." Hannah had a similar experience one day while walking down a street in one of the neighborhoods outside San Francisco when one man confronted her "Go back to your home country!" Sometimes someone will give her the finger. Hannah is African American, "born and raised in America." She has never been abroad.

Yeah, most people think I'm from Sudan. But I have never been there. I have never been out of the country [laughs]. I have had people tell me that I should get back to my country. That I'm not accepted here, you know.

Hannah passes for and is perceived as an "immigrant." Whether in Sweden or the United States, the converts' ethnic and national backgrounds, their belonging, are questioned.

When Layla and her convert friends attended a class she saw the other class participants' difficulties in placing them. Yes, she was light-skinned, she spoke Swedish fluently without any accent, but what about the veil? They searched for an answer in her background.

> When we took this class, it was only during the last semester that anyone dared to ask something like, "by the way, where are your parents from?" "Well, they are Swedish." "Are they Swedish?" They couldn't think that they were Swedish. They knew that we were Swedish [she and her convert friend] and talked perfect Swedish, but there must have been something. Yes, it is strange. There is nobody who thinks that you are entirely Swedish but instead "oh, how well you speak Swedish!"

The women's religious identity is not expected to be an individually chosen one but a consequence of family tradition and background. The women's experiences also reflect how racial and ethnic stereotyping follows certain expectations in regard to language. A veiled Muslim woman is not expected to speak fluent Swedish and, as Marianne pointed out above in the longer interview excerpt, she does not sound like she looks (p. 132). Sometimes people talk *sfi-Swedish* (an abbreviated name of the language classes "Swedish for immigrants") to Cecilia.

> They don't know if I'm a very light-skinned foreigner or a Swede with a veil. So they t-a-l-k l-i-k-e t-h-i-s [she enunciates clearly and slowly] until I start to speak myself and they realize that the chick can speak Swedish. It can be at the post office or people you meet elsewhere. Recently it was at this fair when a friend of mine and I started to talk to some ladies there . . . and my friend is even more light-skinned than me, if possible. And the lady starts to speak s-f-i—S-w-e-d-i-s-h. And we laughed so hard and started to speak. And she just stared and said, "Oh, you know how to speak Swedish!"

Why are the women seen as "traitors to Swedishness"? What are they threatening? Why does their identity as Swedish Muslims provoke people? What kinds of ideas are activated in judgments such as this? Apparently, there are certain ideas about national identity and belonging that are activated and challenged by the conversion to Islam itself. Orvar Löfgren shows how nationality is formed like a secular religion with powerful emotional charge. In Sweden the nation-building during the twentieth century was linked to Social Democracy and the welfare state, "The People's Home." In this process citizenship and cultural homogeneity have been two important components. In the Swedish nation-building, democracy, citizenship, and modernity were three central concepts (Löfgren 1993: 181). Note that these are the same ideas that the converts are accused of betraying since the veil and Muslim identity seem to trigger general associations to fundamentalism, "immigrant cultures," and traditionalism. Furthermore, ideas such as equality, social justice, and solidarity have been fundamental elements in the Swedish national identity for almost half a century (Pred 1998), ideas reproduced in Swedish national self-representations and part of a national discourse. If democracy, social justice, and modernity (but without secularism) are concepts and values the women are accused of being disloyal to, a paradox emerges. These are as we have seen, particularly in the case of Ayşe, ideas that are still quite important.

Many of the converts stress that they are "still the same person," that "you can't ignore who you have been," or that you cannot "take away your roots." While communicating their new Muslim identity to the surrounding world, the women find it as important to simultaneously bring forth their persistent "Swedishness." By using the category "Swedish" the women seek to communicate this salient feeling of self-continuity. We have seen that when Marianne refers to herself as "Swedish," she has particular ideas and attitudes regarding women's rights and the relationship between men and women in mind, ideas that were important even before the conversion. In the encounters with "Arab women" or with some of her husband's relatives, these ideas of hers appear to be of particular importance and significant for who she is. Certain values, formerly taken for granted, become important to confirm and mediate and are given more thought in reflections upon who one is after the conversion.

By accentuating their identity as being *both*, both Muslim and Swedish, they also ventilate a critique of public representations of "Swedishness," of how a Swede should look and act. Their conversion calls for a wider definition of "Swedishness" that includes diverse religious, cultural, and ethnic identities and movements between these (Alsmark 2001). The converts' identities can be understood as examples of a rather new phenomenon in a multicultural Sweden, recognizing not only the cultural diversity of immigrants and refugees but also the commitment of "native Swedes" to a different religious belief such as Islam. In this sense we could understand their visible self-transformation as reflecting main cultural and social transformations. Conversion implies not only a change to but also a change from something. Donning the veil simultaneously as asserting their continued Swedishness and being engaged in women's organizations to ease immigrant women's integration into the Swedish society, they challenge older national notions such as "one nation, one people, one religion," promoting (and themselves embodying) a multicultural society, cultural diversity, and transcultural identities. In that sense they represent alternative versions of Swedishness, a national belonging that is far from fixed or determined, but rather under ongoing negotiations. They are still Swedish but in a different way; they are both Swedish *and* Muslim. To return to the question why the women are seen as betrayers of their national belonging, it is not only their Muslim identity per se that might be seen as threatening but also their simultaneous claim of still being Swedish, unavoidably pushing people to reconsider the meaning of "Swedish" as well as their own Swedish identity.

In the interviews with the American-born converts, strong criticism was ventilated mostly against what they understood as a dominant American culture and politics. Though Fatimah pointed out that she feels very strongly that she is an American, "This is who I am. I was born here, I was raised here" and because it allows her to be both American and Muslim, both Mariam and Hannah ventilated ambivalence and denunciation in regard to their home country. Their conversion urges them to reconsider their national belonging, relating it to their new religious affiliation. In the case of Mariam, the conversion has only reinforced previous feelings of her country and her criticism of "American culture." After her marriage, she and her husband isolated themselves from their family and society in general. For a couple of years the two of them lived in a little cottage in a rather remote area. Initially, they felt that they had to throw away their "American heritage," an idea strongly supported by Islamic discourse that encourages strong ties to *ummah*, the global community of Muslims, rather than to a nation. Mariam explains that she does not feel part of American culture. She does not recognize any American component of herself.

> And I have always considered the American culture as the enemy. You know, I never liked it. I was never part of it as a teenager; I didn't grow up with the kind of pop culture. I have always felt separated from it.

At the time of the interview Mariam was a housewife, running a home-school for the children, and did not seem to interact much with non-Muslims but mostly with people belonging to the Sufi community. Mariam and her family lived in a Muslim country for a couple of years and she has a strong desire to go back. But as she said sighing, her children "feel American" and they do not want to leave the United States. Mariam stressed that she does not feel American. She feels separate from it. In her case her religious identity and some kind of "outsideness" weighs more than any national affiliation. She has the feeling of being "a foreigner everywhere." However, in her reconsideration of national belonging, I would not say that being American is completely irrelevant to her. Being American is important but in a rather negative way; it is a dimension she can not ignore but has to relate her Muslim identity to.

Similar to Fatimah, Hannah resisted any thought that it would be contradictory or fraught with conflict to identify oneself as Muslim simultaneously as, in her case, "African American."[4] Hannah asserts many times that she is proud to be African American; that is who she is. She talks about her background as a more or less inescapable fact, "I'm born that way," but it is not an identity that could prevent her from being something else *too*.

> *Hannah:* I was born American, yeah. When people say, "What are you?" I would say, "I'm an African American." So I was born in America. And they say, "OK, but what are your parents?" A lot of people say that, "you look like you are mixed with something." I say, "I'm sure that there is a mixture somewhere because my father was very light." He had a very light skin. Lighter than I am. I mean, I don't know. It is descended from my ancestors, so I don't know. When I say Muslim I'm not saying that I'm denouncing that I'm an American. I am who I am. I'm an African American Muslim. I always say that. There are some people who have converted and they want to go to another country and live there. My husband wants that too. He wants to go to Morocco. He might go there this summer. He wants me to go too. I might go, I don't really know.
>
> *Anna:* When you say that you are African American, what is it in your background that is important to you? Do you know what I mean?
>
> *Hannah:* You know what? The way I look at that, the way I feel about that is . . . I don't want to feel like I'm ashamed about being an African American. Because I'm not, because that is who I am. I just . . . I think it is not being an African American but the way things are in this country. It is not the fact that I'm African American. I'm proud that I'm African American. I'm born that way. And my family, my family history, we are African Americans. But

I mean . . . I don't want to be another race. You know what I'm saying? I just wished that my family was Muslim. I could have been born Muslim, I could have been raised Muslim. I wish that. I probably shouldn't but I do. That's how I feel. But being African American . . . I think you can be anything you want and be African American. I just don't have to be another race or culture, you know. But I'm proud of that. That is who I am.

I wish I had asked Hannah to further explore her sentiments on this issue. What is she getting at? She stresses, responding to and resisting an internalized public discourse on discrimination of African Americans in the United States, "I don't want to feel like I'm ashamed about being African American. Because I'm not, because that is who I am." "I am who I am. I'm an African American Muslim." There is something that is quite unsettling and problematic to her. Her emotional commitment to Islam makes her wish to have been born and raised Muslim, but while pointing this out she is sure to make it very clear that by saying this she does not mean that she is betraying her background. She has a "solution" to this: "I think you can be anything you want *and* be African American." On the other hand, Hannah expresses ambivalent and negative feelings about her "home country," particularly when it comes to the unjust treatment of herself and other Muslims.

I mean, I don't like America. I live in America so I guess I have to like America. I mean, I don't like the way a lot of Muslims get treated in America. A lot of Muslims go to jail. They [people in the courts] don't always have proof but they suspect. Because they [Muslims] are associated with some political group. As a Muslim I have been through a lot. Being a Muslim is not totally accepted in America. And I mean, it is probably going to get worse. I don't think things are going to get better. My husband always says, he is very political, that in America you as a Muslim are going to get questioned a lot, or stopped a lot in the airports. In the airports I get stopped a lot. My luggage always gets searched. Just because of the way I look and the way I'm dressed.

Hannah talks about how Muslims suffer everyday injustice and discrimination, unequal treatment she knows well as an African American too. Given the practice of veiling, she becomes a possible suspect. During the interviews she is trying to mediate complex ambivalent feelings about her identifications, by employing available labels such as "African American," "American," and "Muslim," over and over. During the course of the interview she seems to associate

"American" with freedom, a freedom that is later distrusted when she talks about the political climate and the situation for Muslims in the United States. She airs a strong ambivalence to her "Americanness" in regard to her home country's attitude to and treatment of the groups that she belongs to. Her "Americanness" gains meaning only in connection with her identity as African American, of which she is proud. On the other hand she is quite ambivalent toward her own personal background. Hannah expresses loyalties to both her African American and Muslim belongings but between the two she senses a gap. Her difficulties lie in the fact that "family" is valued close to her sense of self, but her family members are not *Muslims*. As a response to a question of mine about what it is like to live as an African American Muslim in the United States today, she replies,

> Well, if your family is not Muslim it makes it more difficult. When I'm around Muslims, I feel good: I can't explain the feeling . . . it is a great feeling. I really feel good and I can communicate, you know. I understand. I don't feel different. I don't feel out of place. When I walk around on campus [her workplace] and I see other Muslims, I feel good. We greet each other.

Like Marianne, who feels "100 percent herself" with other converts, Hannah senses a strong feeling of belonging when together with Muslims. Then she feels good. Her Muslim identity is an identity that is of more significance to her sense of self than her American identity. When interacting with Muslims she feels understood. The experience of least tension between selfhood and other people's images and expectations of her is expressed: "I don't feel different." The expression *feeling and not feeling out of place* serves to communicate feelings that are hard to describe. Hannah uses it first in the beginning of the interview to capture her memories of childhood (see chapter 5) when she felt out of place, and later she reuses it to validate her self-transformation and conversion. Only as a Muslim, together with other Muslims, does she not feel out of place.

Today Hannah is a part of a Muslim network of other Muslim families, she is attending Islamic evening classes, she and her family live in an area dominated by Muslims, and her husband works in a school with mostly Muslim children. This sense of community and social belonging with other Muslims strengthens her sense of self as a Muslim. Hannah's insistence on an African American Muslim identity offers to her an alternative version of "Americanness," challenging public ideas and national discourse characterized by materialism, morally declining secularism, and the land of the free.

Becoming Muslim throws the self into a new relationship with one's previous identities and belongings. The convert's experience of being Muslim causes, with necessity, a modified perception of oneself as Swedish or American. Conversion triggers looping effects through which the meaning of national belonging is modified due to the personal meaning given to a Muslim identity.

Chapter Eight

Family, Work, Sisterhood

In this chapter I resume the discussion about the women's various experiences of relationships and how they negotiate their Muslim identity externally and the impact it has on their self-understanding. I begin with a discussion about some of the strategies used by the women to offset widespread assumptions about the "Muslim woman." With particular focus on three significant contexts, that of the interactions with family, employers, and other women and Muslim sisters, this chapter continues to demonstrate the tension between the convert's self-understanding and how she is perceived by others, and between the desire to be recognized as the same and as changed simultaneously. These encounters illustrate the converts as "moving targets," setting in motion looping effects of what it means to be a Muslim woman.

Some Strategies by Targeted Selves

In the previous chapter we saw how sometimes Marianne signals distance and expresses her dissociation from her family members by emphasizing her "Muslimness," her new religious affiliation, solidarity, and ties, and sometimes her "Swedishness" the feeling of being the same. The women employ different strategies to avoid obstacles and to react to harassment and discrimination in their everyday life as Muslims. Particularly in situations when the convert feels questioned and discriminated against, she might employ "defensive practices" (Goffman 1959:13f.) through which she stresses what she is *not*. As a response to stereotypes about Islam and Muslims, the women emphasize counter-images, that they are *not* subordinate to their husbands, that they were *not* forced to convert, that they will *not* force their daughters to wear the veil, that they have *not* embraced their husband's sometimes more traditional understanding and practices of Islam and so forth.

When analyzing strategies there is a problematic aspect that I want to avoid here. It is true that some strategies are merely developed as a response to what they perceive as other people's ideas of them and to discriminatory behavior. But they also have a psychological dimension to them; they mean something to the women. The strategy of Marianne, stressing in front of friends and family that her husband *does* help her in the household, to controvert the idea that she as a Muslim woman could be a suppressed housewife, is a maneuvering with both social and personal meaning. It is not merely rational tactics to appear in better light, to control the impression of the other, to utilize possibilities, but also an act strongly linked to her self-image. On the other hand, we should not neglect dominant messages and power structures behind the strategy that more or less compel her to do it in the first place. She has to relate, in one way or another, to the stereotypes around her.

As Swedish Muslims or American Muslims, the women view themselves as "cultural mediators" between a religious minority and the majority society. This view sets off different kinds of strategies. In the case of the Swedish-born converts, their transcultural identity is perceived as a resource, pleading the Muslims' cause and encouraging a dialogue between native Swedes and immigrated Muslims. Female converts are an important force in the formation of what has been referred to as a "Swedish Islam," initiating and running female organizations, organizing child care, visiting school classes, and having leading positions in Muslim booklets and journals and within the Muslim community in general (Svanberg and Westerlund 1999).

Particularly, Ayşe and Marianne are involved in different activities to counteract prejudice and to alter people's expectations and notions about Islam and Muslims. Sometimes they also visit school classes to talk about the religion and their own conversion. In these social contexts the strategy of showing that one is not as strange and different as people may think is of particular importance.

> You do your best to show that you don't match the stereotypes; so sometimes you maybe exaggerate in the other direction, being very sprightly and funny [laughs]. You show that you are like any other human being, able to laugh and cry. When visiting senior high school the students often ask about sex and such and then you have to give a lot of yourselves. It is good that there are no other Muslims present listening because sometimes it feels like you go too far. You really show them that you are a human being of flesh and blood. That you are not a nun or something.

It becomes crucial to be oneself and to communicate a self-presentation that elicits feelings of recognition and resemblance among the young students. The stereotypical image of the convert not only as a Muslim but also as a religious person in general, to which Ayşe refers above, is made up of ideas that they should be dull and serious, hardly ever laughing. In this particular situation, Ayşe sees it as her mission to under-communicate aspects that could possibly confirm and strengthen negative expectations. She wants to show the human being behind the veil, that she, even as a female Muslim, can stand up and talk casually and gladly about sex. Ayşe continues;

> I just try to be myself. My son plays a lot of hockey and sometimes I go to hockey games together with other altogether ordinary Swedish parents. And we stand there and talk about the game and I feel so normal [laughs] so I forget that I'm wearing the veil. And no one has really asked either. You just show that you are quite an ordinary, normal person. That is the best way.
> *Anna:* I think the veil is in focus for people but after some time it disappears.
> *Ayşe:* Yes, I think so.
> *Anna:* I feel that myself. I must honestly say that in the beginning I focus on the veil, then it disappears.
> *Ayşe:* Yes, and then you see the person behind. I think so too. Especially when you don't make it into something striking yourself. You approach it as natural. "Hi, hi, here I come" and you talk. The veil is becoming rather common in Sweden.

The veil prevents the women from feeling "normal" in some social interactions. And if they do feel "normal" it is because they felt so comfortable that they forgot about their perceived "otherness" for a moment. Many of the converts have got used to the stares, the insults, and the curiosity. In response to public insults they sometime choose to stay quiet, ignoring the comments, even if they cause distress or anger. This strategy of ignoring also reflects the fact that the women want to behave as "good Muslims." However, there were times when the women chose other strategies and responded to the offenses and harassments verbally, like the time when Cecilia was on her way back home on bus no. 4.

> I think people stare at me because I look so obviously Nordic and at the same time I'm wearing the veil. There are those who react simply because of the veil, and become . . . I was on the bus, I was taking number 4 some months ago and then there was this old man who literally sat there snarling at me when I entered the bus.

Anna: What?

Cecilia: He was literally snarling [she mimics his voice] . . . but what noise is that creature making, I was thinking to myself and then I realized that he was sitting there snarling at me.

Anna: How did you know it was toward you? I guess, though, one notices.

Cecilia: Well, because he was sitting looking at me out of the corner of his eyes . . . then I looked at him and gave him the world's greatest smile and said:Boo!

Anna: [Laughs].

Cecilia: He got terrified. After that he was sitting by himself. Then it worked. I thought it was better to do it that way instead of being surly. If I had got surly or reacted and felt uncomfortable, then that would have given him the upper hand. But I laughed at him and was silly and said "boo" and I got in command immediately. So how people relate to you depends on how you relate to yourself.

In the encounter with the old man, Cecilia's understanding of herself as "cocky" is strengthened. Also by narrating this event to me, displaying an example of her decisiveness, she demonstrates and reinforces her self-understanding. Maybe the encounter did make her feel sad and uncomfortable, but even so, she does not show it. In her mind, people relate to you depending on how you relate to yourself. Rather than quietly endure the insolence directed against her, she strikes back. She cannot prevent similar encounters, but with her chosen strategy, responding playfully with a "boo" confident of victory, she tries to shift the power relations and still fashion salient parts of her sense of self as a young, argumentative woman who does not let anyone push her around. In her own words, "attack is the best defense." She feels that she gains control over the unexpected, humiliating situation by employing an unpredicted maneuver, "Boo!" In an improvising move she turns the uncomfortable and unsettling situation into a victory, making a quick recovery from a disparaging position. The man remains quiet during the rest of the bus ride. In response to being classified, Cecilia reacts in an attempt to change the classification of herself.

Hannah told me about a similar verbal response of hers, reflecting a challenge to the stereotype of the veiled Muslim woman as subordinate, passive, and quiet. Now and then she hears comments such as, "Go back to your home country." Well, I am home, she thinks quietly to herself. But sometimes she counteracts.

Actually I have gotten used to it so I don't really react. It might bother me because I'm a human being. Sometimes I say: "May Allah bless you." "May Allah have mercy on you." While some people are really respectful. Sometimes when me and some other women are walking down the street and there is a group of men, they just move out of the way so you can come through. These are the people they perceive as low life or whatever. They give us respect. People who maybe hang out on the street or they may be homeless.

From Hannah's experiences, most of the people who offend her publicly are "white" and the ones who show respect are those she refers to as "low life," poor, and homeless people, people who also have the experience of being socially stigmatized. She too has gotten used to stares and verbal assault. In the encounter, Hannah counters with what has been attacked in the first place, her Muslim belief. Her strategy implies stressing her religious identity, her Muslimness. Then, "May Allah bless you" can be quite an effective and successful reply. The position of defense that these women often find themselves in may also results in a reinforcement of their Muslim identity, even if it is through negative recognition.

Relationship with Family

"I Was Born Christian and I Should Stay that Way"

The title above is a quotation from Hannah. Brought up in a Christian family in the Midwest with a very devout mother and a grandfather who was a preacher, Hannah met some antipathy and lack of understanding after her marriage and conversion. Her mother has had difficulties understanding her daughter's decision, but she has never really explicitly questioned the conversion itself. She is, however, of the opinion that Hannah should stay the way she was born, that is, stay Christian. To maintain a good relationship with her family is essential to Hannah. She often tries to go to the yearly family reunions that are so big that they have to rent a building!

> *Hannah:* My family is Christian so it was like a big shift. My mom is religious. She goes to church, she is really this Christian who believes in going to church.
> *Anna:* So you were raised in a Christian family.
> *Hannah:* Yes, I was raised in a Christian tradition. My grandfather, my mother's father was a minister. So [laughs] all of my relatives are Christians. So it is sort of difficult to be the only one. I get the stares, you know. Everybody looks at me. I remember when I went to a

family reunion in 1995. We have family reunions every year actually, but I don't make it every year but that time I decided to go. And
one of my cousins said, "Why do you wear that rag on your head?"
[Laughs]. And then I have some cousins who really understand.
They say that, "You are in a really great religion. I know a little
about it and I think it is great." But the majority of my family feels
that I was born Christian and I should stay Christian. They think
that I'm brainwashed.
[. . .]
Anna: So your mother never said anything.
Hannah: No, my mother never said anything, not to me anyway. She
asks me questions about the religion but my cousins and aunts are
the ones that question it. My grandfather never really said anything
bad about it. He listened to what I had to say. He never said that
I was wrong. He was the one that I thought would really criticize
since he was a minister [laughs]. But he was really cool about it.
Everybody else they look at me and one of my cousins said, "How
do you stand being a Muslim and married to a Muslim man to
whom you have to tell everything, and report to where you are
going?" Some also ask me about how I should react if my husband
told me that he wanted another wife. This is a question I always get.
[. . .]
My mother lives 3,000 miles from here so I talk to her on the phone.
I went by myself and visited her last December. And what happened
was that I got caught in a snow storm [laughs]. I was delayed two
days when going back home. So I will never go back [laughs]. My
mother and sister come out and visit. My mom is supposed to come
out next summer. I come one year and she comes one year.
Anna: So your family has accepted it?
Hannah: I think they have accepted the fact that I'm a Muslim and
that I dress the way I dress. My mom always asks though why
I don't show my hair. I haven't always worn the scarf. I didn't really
wear the scarf until maybe nine years ago . . . I have worn it back
and forth. But nine years ago I really decided that okay I'm not
going to take my scarf off. But my mom always says, "Why don't
you take your scarf off? Why don't you show your hair? Your hair
is nice." I take it off when I'm with my family and with my husband
and his family but not around any other males that are not family.
Anna: Has your family asked you a lot of questions?
Hannah: Yeah, they have difficulties understanding why I have to
cover my body. I have always dressed modest. I have always covered
my body.

I think Hannah feels some ambivalence about what her family
really thinks about her conversion and Muslim identity. Obviously,

they were not entirely positive about it, it being something new and different. Aunts and cousins questioned her Muslim identity, ventilating prevailing ideas about the oppression of Muslim women and polygamy. These more critical voices have today been replaced by silence. But she also encountered friendly curiosity and acceptance from her grandfather, a minister, and others. Through time, Hannah has experienced a more accepting attitude among her family members. It has now been many years since she converted. They all understand that what they first thought to be only a fad is a prominent part of who she is today.

> *Hannah:* Our relationship is a lot better. The funny thing is that we just don't talk about it. They know that I'm a Muslim; I express it. They don't criticize it anymore and they don't ask questions anymore.
>
> *Anna:* You don't talk about it?
>
> *Hannah:* I mean I bring up stuff but they don't get really into it by asking different questions. Like during Ramadan when I call them and say that we are fasting and I send them a gift, because we exchange gifts like at Christmas time for them when they send Christmas gifts [laughs].
>
> *Anna:* At the same time as you say that family is a very important thing . . .
>
> *Hannah:* Yes, important . . . and I guess I feel blessed that my family is not distant from me because of this. They don't distance themselves from me. They are glad to see me, they call me. We call each other once a month. I don't like celebrating birthdays but every December 28th, that is my birthday, my sister sends me a card [laughs]. So we are not distant from each other. I feel blessed that we are not.

Even if Hannah does not celebrate her birthday, her sister's cards that she receives every birthday mean a lot to her. Despite different worldviews and lifestyles, they are still sisters; they still keep in touch. But their Jane has become Hannah. Just like some of the other women's mothers, Hannah's mother keeps calling her Jane, the name she gave her daughter. She has realized throughout the years that Hannah in many ways still is her Jane, but by insisting on calling her Jane she manifests, time and again, Hannah's background and who she was before, disregarding her conversion. She has probably gone through a period when she feared that her own daughter Jane would become a stranger to her.

Hannah: Even now when I think about my name being Jane, it is a weird feeling. The people who knew me as Jane . . . and sometimes they say, "You know, I call you Hannah because I'm so used to calling you that, but when I think about you you are Jane. And it sometimes feels strange to call you Hannah." But for me to hear the name Jane . . . I still have people in my family that call me Jane.

Anna: How does that feel?

Hannah: They won't accept me for who I am. You know, my new name. I used to get upset [tape ends, change of tape]. So anyway, I mean, I say, "Please call me Hannah." You know, I did change my name." I always do it in a nice way. I never get upset or act crazy or whatever. I just say, "My name is Hannah, can you please call me Hannah even if I realize that you know me as Jane."

Anna: How did your mom react?

Hannah: My mother . . . you know what? My mother still calls me Jane.

Anna: She does?

Hannah: Yeah. I didn't tell you what happened when I went visiting her and I used my new name Hannah. So what happened was that the flight was delayed or something so she was concerned and called the airlines, "Do you have a Jane?" "No, we have no one with that name." Since she didn't really care about my new name, she couldn't remember it. So she had to call my husband and ask, "What is her name?" [Both laugh.] And when I arrived she said, "I really wrote it down. Because if something happens to you I wouldn't know."

Anna: It must be strange for her too.

Hannah: I think it is [laughs].

Anna: Her daughter has become Hannah, right?

Hannah: Right [laughs].

After this incident Hannah's mother will probably not forget that Jane has become Hannah. Most likely it was embarrassing for her to call her son-in-law to ask about her own daughter's new name. These are highly sensitive moments reinforcing an unwanted and distressing gap within their daughter-mother relationship.

Correspondingly, Fatimah grew up in a family where religion was an important and integral part of life. Her parents are committed Catholics and have taken it hard that their daughter has left the religion they taught her during her childhood.

Fatimah: But my family or specially my parents are not so happy that I'm a Muslim. My mother has a hard time with it.

Anna: How did they react when you told them?

Fatimah: They didn't much react when I told them, it is more like little by little they see things that they don't want anything to do with. They don't want to take their shoes off when they are here. My mother makes comments about me wearing my scarf in public.

Anna: She still does that?

Fatimah: Yes, she still does that after eleven, twelve years of it. She still doesn't like if I wear my scarf. And she doesn't understand it. I think my father is more understanding.

Anna: They are religious, practicing Catholics, right?

Fatimah: Yes, so it is hard for them still, especially since they believe the Catholic teaching that everybody else will go to hell. That must be a little scary for them.

Anna: And also that you left what they taught you.

Fatimah: Yes, they say that to me but I don't feel like I left it. I feel like I deepened it.

Anna: What do they say when you say that?

Fatimah: They don't know anything about Islam and they don't really want to. They don't understand.

Anna: They met your husband rather soon after—

Fatimah: Right.

Anna: Was that difficult?

Fatimah: Well, he is a really nice guy, they like him [laughs] but they have a hard time with the fact that he is Muslim and because of him I'm more visibly a Muslim. So they have a hard time with that. But he is such an easygoing person that they can't really find too many faults about him. So I don't think they do.

Particularly Fatimah's mother seems to have difficulties in recognizing and accepting her daughter's choice and new identity. There is also a profound fear caused by their belief as devout Catholics that their daughter will go to hell because she has forsaken the Catholic faith. The scarf signals her daughter's abandonment of her Catholic background and upbringing for a different and "foreign" religion. Fatimah, on the other hand, feels that she can combine her Catholic faith with Islam.

In some ways it is not too hard to understand the parents of Hannah and Fatimah. I think it is important to recognize that their first shock and resistance reflect not merely cultural prejudice but also concern and worry expressed within a parent-child relationship. With time, they seem to have accepted part of it. On the other hand, their daughter's sense of herself as Muslim has not become an integral part in their relationship. The daughter's commitment to something different evokes conflict and fear of estrangement and loss.

As accounts of how they as children have become something fundamentally different to their parents and how their transformation has challenged the parent-child relationship, these are not specific to converts to Islam. I believe there are some congruencies here to the experiences of parents of people who are "coming out" as well as parents of immigrants who fear that their daughter or son will change in the new country and become strangers to them. The convert is becoming something that intrudes upon her parents' memories and sense of who their daughter is. This experienced rupture forces them to rethink in profound ways their own relationship with their child.

"Are You Sure?"

Not all of the women have gone through difficult times with their parents in regard to veiling and the new religious identity. Compared to Fatimah, Cecilia has had a very supportive and understanding father. Her relationship with him is strong and close and she sees him several times a week. Importantly, her father and Cecilia's husband get along very well. Her mother was a bit more reserved in the beginning but she "is adapting to the thought slowly but steadily." Her younger brother, a "cool type," who has friends from diverse ethnic and cultural backgrounds, was supportive, thinking, it was "cool." When she told her parents about her decision, her mother said, "oh, really, well, well," and her father looked at her for a long time and said, "Are you sure?" "Yes," I said. "OK then." Her father had some difficulties at first dealing with all the stares they got when he walked down the streets downtown together with his daughter. But slowly he "discovered that I was the same crazy girl to whom he had been a father his whole life," she said laughing.

> Because it is like I have said the whole time, I'm not thinking of becoming an Arab just because I'm a Muslim. I am who I am. I'm Muslim but I'm still who I am.

I often heard these kinds of assertions about sameness between a "before" and an "after" from the women. The parents' recognition of their daughter is crucial. There is a relief expressed here when parents discover that their daughter did not become the stranger that they expected her to become.

Ayşe told me that her transformation was over such a long period of time that it never caused any shock for her parents. "My parents did, of course, think it was a little strange. Before they realized what it was, they thought that I had joined a strange sect. They see that everything is going well for me so they accept it." When the parents recognize "the same old daughter" they give their approval. Ayşe has also made compromises with her parents by celebrating Christmas every year at her place. Since her parents are not religious, Christmas is not celebrated as a Christian tradition but it is a time when they meet and have a good time together, but without a Christmas tree. Ayşe stressed that she could never tell her parents not to come home to them for Christmas and let them be alone. She knows there are converts who do, but she is not that "extreme." Ayşe emphasized, "It is not in my nature."

Her parents have also made a trip to Ayşe's husband's home country, visiting their grandson who was there studying Arabic. They were then in their seventies and it was "the trip of their life." It was a fascinating experience to them. They met their son-in-law's relatives but lived in a hotel so they could eat "beef steak and French fries." The trip had an impact on their attitude toward their daughter's conversion. By gaining personal experiences of their son-in-law's country and culture, and therefore also of a part of their daughter's Muslim identity, it improved their understanding of her choice to convert. However, just like Hannah's mother, her parents still call her Malin. Ayşe asserts that even if her name is Malin Ayşe today, she is not any less Swedish than before. "I'm still Malin . . . one is still the same person. I feel Swedish in the first place; one is Swedish. It is not really such a transformation." Thus, not only the parents' approval of the decision but also their very recognition of their daughter, that she has not totally changed after all, is crucial both for the covert and her parents, reinforcing a sense of continuity with one's past as well as one's earlier relationship with family members.

At Work—Visibility and Vulnerability in Public

The Veil

The veil stands for so much. It stands for suppression. The symbol for the suppression, to force something under something. The underlying, the hidden. There are so much symbolic values around the veil. If I had taken it off no living soul would care if I was Muslim.

Marianne formulates here what she sees as the problematic aspect of the veil. The meaning it has for her does not correspond to common public understandings of it as a marker of religious repression and fanaticism, male oppression of women, and traditionalism. At work the converts' experiences of their employers' or customers' reactions to them point to one and only one thing. In these encounters it is all about the veil.

Hijab became a hot controversial question in both Sweden and Denmark when several chain stores, such as *Åhlens* and *Magasin du Nord*, forbade their staff to wear Muslim coverings since it breaks with the requirement of uniform dress code. A report from the College of Malmö indicates that thirteen out of fifteen employers stated that if the applicant wore a scarf it would affect their decision to employ her.[1] However, the employers do not understand the restriction as an expression of religious discrimination but as something that is applicable to all kinds of divergent appearances, comparable to nose rings or piercing in the tongue. In a conducted public opinion poll from 1999, 56 percent of those questioned were of the opinion that employers have the right to forbid the Muslim *hijab*. An interesting result from the same poll shows that 67 percent thought that the Muslim girls should have the right to wear the veil at school.[2] The veil has also become an effective tool in arguments for or against "multicultural Sweden" in public debates. In August 1998 a letter was sent to the editor in a local newspaper with the title "The veil made her even more beautiful," in which the writer described the veil as decorative jewelry against the "the golden-brown color of the face."[3] The chairman of the immigration association of the Skåne Party (*Skånepartiet*, which is an anti-immigration, right-wing party) replied that "to be able to prevent the dark fundamental forces it is important to support the Skåne Party's proposal of a local prohibition against wearing the women suppressive Islamic veils."[4]

More recently the French government passed a law banning the Islamic headscarf, together with *kippot*, the Jewish skullcaps, Sikh turbans, and large Christian crucifixes, in state schools. In February 2004 the French National Assembly voted overwhelmingly in favor of the law with arguments resting on a main French principle, the secular state. In the United States there have also been several lawsuits concerning teachers' right or not to wear the *hijab* in their profession (Moore 1998). The *hijab* has caused a heated debate, fueled by present political discourses on Islamic terrorism and segregation. Arguments based on the neutral, secular state stand against the constitutional right of religious freedom and antidiscrimination law.

Despite experienced difficulties, all of the women have adopted in various ways the Muslim tradition and regulation of veiling. The decision to wear the veil was not easy, causing anxiety and stress, but it also concreted everyday obstacles in different encounters. By donning the veil, the convert can cover her appearance but she cannot cover her identity and affiliations, quite the opposite. This is the important communicative aspect of the veil. She may not perceive herself as being an object of men's gaze and temptation any longer but she is, as discussed earlier, an object for people's crude categorization. As Marianne commented above, "If I had taken it off no living soul would care if I was Muslim." However, converting to Islam but not wearing the veil and then consequently not getting any reaction to or confirmation of the change (at least in her encounters on the street, in the post office, and with others who do not know her) could be at least as difficult, albeit in another way. With the veil she will endure discrimination and harassment but, on the other hand, this negative confirmation confirms the change she has undergone and might even strengthen her Muslim identity.

Encounters as Struggles

Fatimah: So for me it [the veil] is a part of who I am now. It is not something that I look on as a burden. And again, even in the struggle that I'm in right now, going out into these offices, part of what it is, is that when I walk into the office I want them to like me. I don't want them to judge me based on the fact that I wear the scarf. They should have a thirty-second encounter, not sit down and have an hour-long talk to find out who I am. The veil is such a very quick visible thing. And I have to make a decision in that.

Anna: You mean that a lot of time goes to questions?

Fatimah: Yes, it is their first impression, you know. And are they going to call us back or are they not? I'm struggling with that. And so far I have been wearing my scarf anyway. Although a couple of times I have taken it off particularly when I know that I'm dealing with older people. If I'm going to deliver a report to an older person, I don't want them to be freaked out by it. I just want them to get their inspection report and read it.

Anna: You take it off in the car?

Fatimah: Yes, I take it off in the car and go in and come back and put it on. But for the most part I just wear it anyway. It is funny, that has been more of a struggle and I haven't had to deal with it for a long time.

Fatimah and her husband run their own business, and during the last year Fatimah has started to help out more in the office. Before, as a stay-at-home mom, she had been rather protected from these kinds of public interactions and people's reactions to her clothing. In the immediate encounter at work, on seeing the veil, the customers suddenly become very interested in finding out *who* Fatimah is. Instead of focusing on the inspection or the report, the reason for their visit, people are curious to hear her story. What is she? Why did she convert? Why does she cover her hair? Instead of having an interaction and communication at a strict business level, Fatimah is put on the spot to explain, clarify, and try to make people understand. She feels more or less obliged to let strangers into a personal space. Their roles, consultant and customer, are overshadowed by questions regarding her personal identity and story. Suddenly an anonymous and professional meeting turns into a one-way inquiry into private matters. Fatimah experiences these encounters as "struggles"; each a struggle to make them like her despite any possible negative associations they might make in the first contact. "I want them to like me. I don't want them to judge me based on the fact that I wear the scarf." It is a struggle between being able to openly be who she is and at the same time running the business well, not scaring away any customers. Afraid of freaking somebody out, Fatimah sometimes takes off the veil in the car particularly when delivering a report to an older person. Fatimah ventilates here a human desire to be liked, recognized, and accepted for being the one she perceives herself to be. This involves a struggle to communicate to others what it means to her to be Muslim.

"I'll Call the Discrimination Committee!"

One morning Cecilia steps into her employer's office with the veil on for the first time. She tells him that she has converted to Islam and that she is going to wear the cover during working hours.

> And then I came walking to my work and "hi." And he [the employer] just stood there staring at me and then he looked out the window and it was not cold outside, so he started to realize that the veil was maybe not a covering that I wore because it was cold out but for another reason. "Do you know that I will wear this from now on when I work here?" I said in a saucy way. It is better to attack than to ask. "Well, I don't know." And then he started to talk about the firm's business and so on, this way and that way . . . But I said, "I know there are other people that have got permission to wear a veil. I will have the veil on and if that doesn't work, I will go and talk to the discrimination committee." "Hmmm, OK then, but none of those big sheets." Oh dear! "And none of those screaming colors because it should

harmonize with the working-clothes." And I guess I can understand that. But he doesn't like it at all. Not at all, but he has to accept it.

There is a great variation between the women's experiences of and strategies employed toward employers. It depends not only on the type of work and the employer, but also on what kinds of attitudes and tactics the women employ themselves. Cecilia understands herself as insolent and fearless and she approaches her employer as such. She gives a self-confident expression and she makes sure the employer knows that she is aware of her rights. Her employer did not like it at all, but Cecilia's reference to the discrimination committee worked well. Even if it is often overlooked, there is a law against this type of discrimination.[5] The employer accepted it as long as it was not too eye-catching a color. Her working colleagues, who are both "Swedes" and "Muslims," have not said or commented much on the change of clothing. The Muslim men "have always been respectful but now they are almost reverent. And the women just think it is cool."

"I Employed You as an Ordinary Swede"

It has been quite different for Ayşe. Unlike Cecilia, she encountered strong resistance to her assuming a religious identity at work. There is a prevalent view of religion in Sweden that it is something "private" and should be practiced at home. Expressing your religious affiliations publicly often seems to aggravate strong opposition and emotions.

So when I converted I went to the boss and told him that I had converted to Islam and that I would very much appreciate if I could dress as a Muslim. But that wasn't really popular. "No, that won't go. You will scare away the customers and no, you can't." "Oh well," I said, "I guess I have to find myself a new job then." But I had to have a job so I could earn my living so I stayed some more years.

Ayşe's employer expresses fear not only of the different forms of visible religious commitment but also particularly of economic loss. There is another argument that Ayşe heard: "I didn't employ you like that, but as an ordinary Swede."[6] The employer gives her two alternatives, either to stay as she was before, that is, as a "normal" Swede, or quit her job. For a long period, Ayşe chose for financial reasons to keep her work, not wearing the veil. Like Fatimah, she took it off in the car in the garage before each time she took on her professional role. After work she put it on again in the car. With this kind of strategy they avoid possible negative reactions and conflicts, but on the other hand, an essential part of their identity is denied.

During these years Ayşe lived a double life. At work she passed as what her employer called an "ordinary Swede" and after working hours Ayşe again "became" a visible Muslim. Obviously, for her employer an "ordinary Swede" does not wear a veil. By stressing her cultural difference and divergence it seems as if the discrimination and special treatment become legitimized. Interestingly, some of the women, like Ayşe, have later chosen another way of still having a professional job without having to deal with encounters with critical non-Muslims questioning their identity. Some of the converts run a cooperative business, in which the veil is neither questioned nor makes them stick out as problematically "different."

Hannah—I Love My Job!

This college is more diverse and I feel more comfortable than I have ever done before in a job. We don't celebrate holidays. There are too many people with different backgrounds and then too many holidays to celebrate [laughs]. I love my job! I can be myself and I feel comfortable there. People look at me as the one I am. They respect my religion. I'm happy about that. And there are other women that are Muslim.

At the time of the interview Hannah was working at a childcare center at a multicultural college. Here she encounters tolerance and recognition of her identity as a Muslim among many other ethnic and religious groups. She does not need to either over- or undercommunicate her Muslimness. In the context of diversity she feels "comfortable." I could tell that she was at ease during one of our conversations that took place at the schoolyard when people, Muslims as well as non-Muslims, were passing by, greeting her.

It was during our conversation about the veil and the obstacles that follow it that Hannah started to talk about some difficult job interviews.

I was worried what would happen if I went to a job interview. A couple of times when I have gone to job interviews, I have felt that it has mattered. One interview when I walked in, I mean they haven't seen me before and when I stepped in you can see it on their mouths. She asked me a lot of tough questions that she didn't have to ask. I think she did it because she was trying to . . . I don't know, I just felt that way. The reason why I say this is because another girl, who is Hindu, Indian, she went to the same job interview and she felt the same way. I don't want to say that they are prejudiced but I felt that way. I don't have any proof and I'm not allowed to say things I don't know are true. But I really felt that way.

Anna: What was the job?

Hannah: It was working as a childcare . . . it was being in charge of the program. I guess they didn't want either her or me in charge. It was mostly white children and some Asian and maybe one or two African American children and one Hispanic child. The majority was white. I never know how people will perceive me because the way I'm dressed.

As a good Muslim, Hannah does not want to accuse the job interviewer of discrimination but she as well as her friend are quite convinced that their "different" religious and ethnic backgrounds from that of the white kids that they would be in charge of mattered in the decision not to employ them. However, the object of scrutiny is possibly not only Hannah's religious identity but also her "race," her being African American.

These are uncomfortable, discriminatory situations, triggering sensitivity regarding their identity and reinforcing a sense of outsideness. This uneasiness is what the converts want to spare other people as well as themselves when they take their veil off or "warn" people ahead of time that they are Muslims. In the public workplace and in their interaction with employers and customers, their Muslim belonging is once again put in the spotlight. Some of the women find themselves in a working environment that accepts their identity and in which they can move with ease, while others encounter intolerance and find themselves denying a part of themselves, by leaving the veil in the car.

New Womanhood and Female Encounters

One dimension of the women's conversion reflects a desire to find new ways of negotiating and exploring their identities as women. As previously discussed, the conversion allows for the convert to elaborate and try out not only a new religious self but also a sense of new femininity, a new gendered self. This newly sensed femininity, which partly involves criticism against Western and Muslim patriarchal ideals and practices as well as identification with a Muslim sisterhood, is implemented and reinforced in their everyday life. Interacting with other veiled, female Muslims is an important part of their self-making as Muslim women. In relation to other Muslim "sisters" they confirm their Muslim identity and explore femininities that might also be expected of them.

Some of the Swedish-born women elaborated on how the conversion and their new roles as cultural mediators, giving talks in schools, have extended their social network considerably and brought with it an increased self-confidence. As Muslims they have gained not only an existential meaning in their life but also a new social role and function in the multicultural Sweden. Certain interactions and practices reinforce this new self-understanding. Their insistence on their continued "Swedishness" partly reflects their strongly emphasized support and commitment to women's issues, albeit explained through new texts and social belongings. As an explicit resistance to stereotyped ideas that their conversion of necessity should have brought with it oppression, they stress their feminist arguments as well as a combination of seemingly two separate representations of gender roles.

Marianne, Ayşe, Zarah, and Layla are all involved in different Muslim women's organizations and activities. They have traveled to countries all over the world attending international women's conferences and meetings. While describing events where Muslim women from different parts of the world have come together, the women often become quite emotionally engaged. These new ways of interacting with other women in gender-segregated settings have acquired salient meaning important to their identity-making as Muslim women. In these different kinds of "women encounters" a new gendered self-image is tried out. Also, they allow them to experience a global sisterhood, a sense of solidarity ideally beyond national, racial, and ethnic borders. Ayşe told me about a conference in Africa that she had attended together with some of her convert friends.

> Especially when we were at this women's conference; it was very *wow*, what self-confidence one got there! One saw all these strong women from all over the world, professors, physicians, and lawyers, super-women, that is. Strong. Oh, it felt like we could take on the whole world when we were leaving. We were treated very nicely at the hotel, we were driven around in cars with police sirens and receptions and hotel gardens, all that you know. It was totally . . . cocktails without alcohol. It was totally exceptional.

The traveling, attending conferences, and involvement in different women organizations have implied a whole set of new practices and interpersonal relationships. Interacting with other veiled women, talking about gender equality and the different ways of approaching gender issues in Islamic texts, and how they can be implemented in a modern society, makes them feel part of something of universal importance. Ayşe expresses a strong admiration for the well-educated and knowledgeable women; with them she identifies herself as a Muslim woman. When she enters into this new female world she acquires a powerful feeling of being different from other Swedish women, a positively sensed difference that connects her to women from many different cultures. If she had never converted to Islam, Ayşe believes that she would have lived a dull life as a "middle-Svensson" (an average Swede).

In a fascinating way, questions of selfhood, womanhood, and sisterhood emerge in these contexts. The new experiences and practices within a Muslim female world have changed her own self-understanding; she has gained a stronger self-assurance. Furthermore, Ayşe's *personal model of social justice and solidarity with the Third World* is acted out and becomes invigorated through the strong community of women from all over the world and the comradeship within it, regardless of national, ethnic, or professional background. Likewise, Cecilia told me about a wedding party that she had attended recently with only women. Occasions such as these offer a social opportunity for the women to unveil, to dance, and act in ways that is not allowed or accepted in other situations. Without the veil, Cecilia and her friends had, ironically, even passed the bride not recognizing her and vice versa.

> Cecilia: Of course it becomes more relaxed with only women, with sisters, you know, you share a common worldview and goals in life. Obviously, you can relax more. It is actually quite fun. We were at a wedding party with some sisters some months ago. Then all of the women that you otherwise are used to seeing with *hijab* take it off and walk around without it and then we didn't really recognize each other. Wow, in regular T-shirts and dresses. Then it was a party!
> Anna: You celebrated with only women?
> Cecilia: Hmm, and then it was really a party.
> Anna: And no one wore the veil?
> Cecilia: No, we danced, ate, and laughed and talked rubbish. We even had hired a belly dancer [laughs]. These things we would never have been able to do with even one man present. But it was forbidden to have cameras and video cameras there. Everyone could relax and there was a totally different atmosphere.

It is with female Muslims, and particularly with converts, that the women feel that they can be themselves, not having to worry about what people will think, and it is to these social relations and cultural activities, such as conferences, summer camps, wedding parties, and reading groups, that they bring certain personal feelings, desires, and thoughts. While socializing with women, with whom they share a profound experience and common belief, they can test, compare, and reflect upon ideas, even if they differ in their views.

At "home," however, they feel often criticized by middle-age women who consider themselves feminists and believe that the converts have betrayed women's rights. Lisa told me about a particular incident when she was visiting a school class to talk about Islam and her conversion. During the visit, she felt that the teacher verbally attacked her, accusing her of being submissive, letting her husband dominate her. In the encounter a struggle between different definitions of freedom and rights was manifested. Instead of letting Lisa talk about her own experiences as a Muslim woman, the teacher offended and criticized her openly in front of the class.

> It was this woman who talked about freedom and that the veil is a sign of letting the men dominate us women. It was during a visit to a school where I was going to give a talk and the teacher . . . I was thinking, excuse me, am I not here for the students? Not for the purpose of having you vent all your prejudices. That was what I was thinking.
>
> *Anna:* You were there to talk and the teacher told you this?
>
> *Lisa:* She herself was just about the age of fifty. And she had a skirt that was very short and with a slit. Here one can laugh at those who wear that kind of dress and that are too old for it. There is a lot of oppression here . . . if you are not good-looking enough to wear such a skirt you are perceived as ridiculous instead.
>
> *Anna:* But what did she say?
>
> *Lisa:* She said, "Do you wear that [the veil] to show that the men dominate and that you are subordinate?"
>
> *Anna:* What did you tell her then?
>
> *Lisa:* No, it is not the men you should subordinate to but to God.
>
> *Anna:* And that answer is maybe not always easy to perceive either for a Swedish woman.
>
> *Lisa:* No, that annoys them even more. But that is pretty clear, how should one explain this to a person who does not even believe in God? But on the other hand, no one can really argue against it because if I say that I believe in God and that I believe that God wants me to do this, there is no one that can . . . OK, people can think that I'm nuts. But they cannot argue against it. So in some ways it is a rather unbeatable argument.

Anna: What did she say after that?

Lisa: Well, she didn't say much. I believe that God exists and that he has said this, *the end*. But the young girls in the class said that it could be good [to wear the modest Muslim dress] so that the men don't think dirty thoughts.

Anna: How old were they?

Lisa: Sixteen, seventeen, maybe. I mean, all children and young people understand pretty easily. These problems appear higher up in the ages.

Anna: Why do you think the young ones understand better?

Lisa: They have probably not as many prejudices. They are actually pretty open-minded. They can still take in different things. I think younger people understand these things regarding the spiritual better. In the movement of '68, not only ideas about the women's liberation are included, but also atheism. It is a package of ideas that is resistant to Islam in general. It is about communism and atheism and a particular perception about the meaning of freedom.

Lisa is here talking about ideas regarding women's gained freedom, atheism, and socialism, ideas that she refers to as part of "the movement of '68." She believes these are strongly held ideas among middle-age women, who were young during this period and who have experienced increased equality between men and women with improved possibilities on the job market in comparison to their mothers. She also briefly connects these thoughts to socialism and secularism and a negative feeling toward religion in general. Paradoxically, these criticized middle-age women often view themselves as open-minded and liberal. Here we can detect a sharp criticism against traditional Western feminism, which has too often approached Muslim women, and "Third World women" in general, as subordinate and docile victims, neglecting, if not even disapproving, the role of religion in women's lives (Mohanty 2003, Narayan 1997).

Marianne expressed similar ideas when talking about a general notion of her as a Swedish Muslim woman. She believes that many people view the woman who voluntarily converts to Islam as giving up her rights, the rights that required a long history of feminist struggle.

I think I signal a provocation against their ideal of equality. That is my interpretation of what they think. But maybe they don't. But I feel like they think that I'm a reactionary type who has given up all the goals of the women's liberation; that I'm conservative and that I have accepted the ways men look upon a woman.

I believe some Western feminists, as well as others, have difficulties viewing the veil as not being a symbol of oppression of women and a hindrance to the liberation of women.[7] It is also interesting to note that similar criticism is directed against other "Oriental" practices, such as Oriental dance or "belly dance," as it is often referred to in the West. Even if it is a very different and a much more "revealing" practice compared to the veiling, it is similarly targeted in the West as an expression of patriarchal oppression of the woman, reducing her to an object. However, Oriental dance in Sweden can also be seen as offering alternative experiences of femininity. Similarly to the converts, some "belly dancers" are actively engaged in women's rights (Berg 2001:203). Veiling, like Oriental dance, composes different ways of exploring other female identities, while simultaneously expressing criticism of "Western" notions and practices.

Feminism is, as many feminist researchers have pointed out, no longer the cause of exclusively white, middle-class Western women. In a postcolonial world there are several different definitions of feminism and movements fighting for women's rights and liberation emerging in the so-called Third World as well as among different minority groups in the West (Bulbeck 1998). The women I have met may represent a new phenomenon in their home countries in which new definitions of gender relations are made drawing from other cultural representations.

It is not only Western women who attack the converts for veiling but also some "Muslim" women who have immigrated to Sweden. The converts sometimes encounter women from Muslim countries who are secular and very critical of Islam and the treatment of women in Muslim countries. The converts share some of these opinions, but the encounters can still become quite problematic. With their own experiences of limited freedom and violent sanctions against women in their Muslim home countries, these women have difficulties understanding how any Swedish woman at all can deliberately choose to put on a veil. For them the veil symbolizes oppressive regimes and reminds them of lived difficulties. Cecilia told me about her Iranian woman friend and her strong emotional reaction when Cecilia converted.

> This Iranian woman just jumped on me totally hysterical and was close to scratching my eyes out. Then I told her, "Listen, Sweden is a democracy, in Iran you are forced to wear the veil but in Sweden we have democracy and therefore I have the right to wear it if I *want*." She can't tell me not to wear it just because she doesn't want it in Iran.

There are many Iranians who say that they think that one should forbid the veil in Sweden on the whole. But then one does the same thing as in Iran but the opposite . . . Here you are allowed to go without if you want but you are also allowed to choose to wear it if you want.

The Swedish women's identity-making as Muslim implies a strong insistence on their continued feminist views and commitment to women's rights, a commitment they have to defend both with women from Muslim and Western countries. We have to move beyond the simple dichotomy of West/East, individualism/collectivism, and modern/traditional, where religion represents the past, tradition, and timelessness while modernity represents reason, rationality, and change. Islam is not a homogenous and monolithic whole. Just like other systems of ideas, public representations, or ideologies, it takes on different cultural and idiosyncratic forms and expressions depending on the context and personal appropriations and backgrounds.

The formation of alternative gendered subjectivities is manifested and reinforced in the social interactions of everyday life, in which the women engage some representations while they resist different expectations and classifications of them, and by doing so they make room for a new way of defining themselves. Through social practices, such as organizing international women's meetings, starting women's organizations to support and help the integration of Muslim immigrant women and kindergartens for Muslim kids, and also visiting school classes and being interviewed by journalists and researchers, they try out an identity as Swedish Muslim woman, challenging public representations of the "Muslim woman."

The Other?

In this last part of the book different kinds of encounters are discussed, in which the women's Muslim identity is in the center of attention evoking a whole range of emotional and prejudiced reactions. The mothers of Hannah and Fatimah have difficulties acknowledging and understanding their daughter's choice and its consequences, Ayşe's employer warned her that if she did not revert to the ordinary Swede she once was he would fire her, and one day a young lad shouted to her "Go home you fucking Arab hag!" Why these common strong reactions? Why is the women's conversion so provocative? I believe the provocation, the seemingly perceived threat, partly lies in the women's very *deliberate* choice to convert to Islam. In addition, the view of them as "traitors" reflects a common thought of conversion as an absolute change entailing a total denial of who one was before and the idea that it is not possible to be both, say, Swedish and Muslim simultaneously.

In an article a Swedish researcher offers the reader a little experiment of thought, highly aware that the subject has been discussed to death: "Imagine that your daughter tells you that she is going to marry a Muslim and/or convert to Islam. How would you react? And why, based on what assumptions, would you react the way you would? What would be the reasons and motives of your reaction?" (Sander 1995:161f., my translation). In regard to the women's experiences as well as the frequent upset responses I have received myself when talking about my research, I believe that many parents would probably have, at least initially, some kind of disapproving reaction, going through feelings of betrayal, anger, concerns, uneasiness, and fear. I also believe that the responses to the question "why?" touch upon the notion of cultural differences and widely spread judgmental images of Muslims and Islam. However, as we have seen, some parents have been more approving and supporting after seeing that their daughter is content with life and doing well and has not, as expected, become a totally different person. The emotional difficulties seem to concern what the parents experience as their daughter's denial of her background, and then implicitly of them as parents, and the fear of separation. The *recognition* of some aspect of their daughter's personality is significant.

Obviously, the encounters excite rather strong emotions, perhaps sometimes hard to explain, toward what may be perceived as "foreign" and "different." Sander proposes that the attitudes and explanations to the questions above "often lie within the following 'obscure' and in a sense metaphysical area: mixture is experienced as 'bad,' often linked to phenomena such as impurity and the like, while purity of different kinds is experienced as 'good' " (ibid.:162, my translation, see also Douglas 1966). As Marianne pointed out, she thinks that people have difficulties in categorizing her culturally, which may lead to uncertainty and confusion. The converts' transcultural identity, and frequent insistence on being both, could be experienced as problematic since it questions a perception of cultures as isolated and stable entities and that they should stay that way. And, according to this way of thinking, a mixture would unavoidably lead to confrontation. From this perspective, the converts are viewed as anomalies, "which are both neither-nor and both-and: which cross borders of categories" (Eriksen 1994:51, my translation). Sometimes this manifests itself when ethnic stereotyping leads to racial effects and discrimination. That was clearly the case when someone shouted on the street to Ayşe: "Go home, you fucking Arab hag!"

Aversion against Islam and negative attitudes toward Muslims in Western societies are often explained through the Western historical and cultural construction of the Muslim as "the Other," the mirror image of what we, "Westerners," are not. This is and has been an interesting and fruitful perspective, particularly by revealing historical power structures and relations between West and East (Said 1978). But it seems to me that it reaches a certain level of analysis but goes no further. Many phenomena around us, gender as well as national and racial constructs can be explained by the construction of the other's otherness, the construction of the inferiority, backwardness, and barbarism of "them" versus the superiority, modernity, and civilization of "us." However, when placed in the personal world of the convert, social-construction talk about her as "the other" becomes tricky. The thesis offers comprehensive explanations for general attitudes and dominant political discourses and interpretations as to why they take these expressions, which are significant questions indeed. But still, we are left with the question of what consequences "the talk of the Other" has on "the Other." Namely, the inevitable looping effects that take place when persons interact with and engage public classifications.

The cultural constructed "Other" radically different from "Us" is an invention of, and has long been a project in, anthropology. The view of culture as an archipelago of isolated islands nourishes the evocation of this otherness. Today researchers and journalists explain problems in the multicultural society and prevalent stereotypes of Muslims as a product and reflection of this Western constructed thought. Muslims are seen as "the Other," no longer far away "in the Orient," but next door. Again, this is important in pointing to persistent historical structures underlying social thought, public representations, and politics, but what impact does it have if the analysis stays on this level? An unfortunate, though not intended, consequence could be that we also assume that Muslims do feel as if they embody a radical otherness. The Orientalist discourse (constructed by an intellectual elite) is then assumed to have powerful impact on the mind of "the Other" (Ewing 1997). In the convert's identity-making, resistance is put up not only against this discourse, the classifications of her as "an Other," but also against other social classifications. Instead of assuming, a significant question that we need to ask is whether this discourse does have an effect on the individual in her/his everyday life and, if so, how. Do the encounters bring with them feelings of being "the Other" for the women? Not necessarily. By assuming both the dominance of the discourse over people's judgment and its impact on the subject, in this case the female converts, I believe one risks supporting and reinforcing a discourse that one is trying to criticize and analyze in the first place.

The women expressed how they sometimes find themselves being perceived and treated as "odd," "foreign," "different," and "not normal." The experiences of different sensitive relationships and social interactions, such as parents' difficulties accepting their daughter's religious identity, employers' discriminating comments, and anonymous people's disrespect on the bus ride home or telling them to go back to their home country, do something to the self. Their identities as "Swedish Muslim," "African American Muslim," or "American Muslim" engender an often experienced gap between their own sense of who they are and others' perceptions of them, but not necessarily an experienced and internalized "otherness." The women may be viewed as "the Other" by some, but they are not passive recipients of cultural labels. When the women are treated as the "Other," they talk back, eliciting looping effects of the meaning of the category of the "Other." The tension between self and others spurs a heightened awareness of the self and the very aspiration to reconcile the uneasy disparity by conveying *their* personal meaning of being Muslim. The encounters, the interviews included, trigger an acute awareness of a targeted selfhood and a recognition of their self-transformation. But equally, as important as displaying and implementing the changes brought about by the conversion and having this transformation acknowledged, is the convert's sense of continuity, her feeling of still being the "same" rather than a changed "other," and having this sameness recognized.

Chapter Nine

Personal Versions of Islam

The "becoming" in the title of the book points to a crucial theme in the study—that of process. Becoming points to a happening, something turning into something. The conversion, the process of becoming Muslim, is neither final nor predictable. When the self is becoming there are no sudden breaks or absolute changes, it is gradual without any fixed points. One day is like the other, but still not. Just like the converts, we have to negotiate our identities continuously throughout life in regard to new experiences and events. In this last chapter I want to conclude with some encapsulating thoughts and observations.

Throughout the book we have witnessed the ongoing process of meaning-making, how the converts are making sense of the conversion, changes in identity and life, as well as the classifications of them. The conversion itself prompts careful considerations of who one is, where one is heading, and what various categories of belonging mean. This meaning-making is brought about by a continuous interaction between the self and public representations, between mental processes and external messages.

Meanings are neither given nor static, but shifting and transitional, and are assigned by individuals. This aspect of meaning is discernable in how the women negotiate their identity both internally and externally. When the women appropriate Islam through dynamic personal models, making it into their own, as well as responding to the categories placed upon them in interpersonal interactions, looping effects of meaning are spurred. The least we can do then is to attend to the personal worlds of the women, and that is what I have aspired to in this book.

There is a prevalent idea, both as a general notion and within academic discourse, that when people go through transformations, be it religious or political conversion or some kind of rite of passage, they would by necessity become profoundly different persons, adjusting fully to a new ideology or taking on a new social role and leaving behind an old self. This view rests on an assumption of the power of language, ideologies, and public representations that when a person internalizes a belief system it is taken in as a whole. But as I have attempted to show, the self is not a passive construction of external messages. My work addresses thus an ongoing debate about the construction of identity and urges anthropologists, and social scientists in general, to re-center the self in their projects in order to attend to some central questions: How do public messages and symbols become meaningful? In what way can we explain how people create self-coherence and sense of continuity despite the appropriation of multiple and seemingly irreconcilable discourses?

A central claim in the book is that conversion is impelled by personal models, idiosyncratic patterns of emotionally salient personal ideas and experiences about oneself and one's place in the world. It is these personal models, rather than public discourses, that incite and guide the women's conversion. The analysis of these models allows for the exploration of the women's own understandings of the religion, their very personal versions of Islam. The converts evidently share many beliefs with each other; there is a consensus regarding the basic principles constituting Islam; this consensus allows them to identify as Muslims collectively. But besides this "social sharedness" of ideas, Islam means different things to the women. Islam is infused with unique personal experiences and memories that give salience to the belief and make it meaningful to identify as Muslim. Mariam talked about a spiritual connection, traced back to childhood memories of her father telling stories about the angels and Mary; Fatimah expressed an existential anxiety and her spiritual quest for meaning; Lisa referred to a personal crisis and a strong inner desire to feel a connection with God; Hannah repeatedly stressed the salience of strong family ties with the background of how she as a child witnessed betrayal and cheating; Marianne explored her commitment to women's rights and expressed, like Cecilia, a strong longing for a change of self. These are personal quests in particular biographies.

Becoming Muslim, or as Cecilia expressed it, the step toward something unknown, has personal resonance. The women expressed a desire and a longing for change in life, a quest sometimes inspired by traveling to and living in a far-away place, a personal crisis, or experienced lack of spiritual meaning in life. But a conversion, and a sense of change, inner as well as outer, does not happen automatically. The conversion requires reorganization of personal identity and of life story, as well as the doings in everyday life. Mariam and her husband chose to live in an isolated cabin far away from external influence in order to carry out their inner belief.

> Because we initially felt that when you become a Muslim, you are throwing away your American heritage and your family ties. And I think that lot of families feel that way, betrayed, due to this. You feel that everything in your life has to be different. You are not the same person and if you hang out in the same place you won't change. You have to throw everything away.
> *Anna:* You say you are not the same person anymore—
> *Mariam:* Well, I don't think that is true. But I think . . . in order to implement the changes you *have to* glamorize it, exaggerate it to that extent. To make it . . . it takes a while to make anything natural.

Just as Mariam asserts, it takes time to make something "natural"; it takes time to assimilate a new religious belief. Becoming Muslim does not imply a fundamental shift when the person suddenly becomes a believer. Rather it suggests a process of reconciliation of ideas and trying out a new identity.

The women implement the conversion through many different means such as veiling, change of social network and friends, praying, and change of career plans. But besides the changes in lifestyle and self-identity there is an unyielding sense of being the same. The tension between past and present, between feelings of being the same and being different, is poignant in the women's life experiences. In the quote above when I call into attention Mariam's own previous expression that one is "not the same person" anymore she replies immediately that that is not really true. She knows that she has not become someone else; she has changed but she is still Mariam, an experience articulated by many of the women, reflected in statements such as "I'm still the same," "I am who I am, I'm Muslim but I'm still who I am."

How do people change? How can we explain the convert's concurrent feelings of change and continuity, of being the same and being different? What is changing and what is staying the same in their experiences? I have argued that transformation, as well as the feeling of permanence, is made possible through cognitive recognition and reconciliation.

Throughout the book it has been demonstrated that Islam appeals and becomes meaningful to the women since the Muslim belief and practices address and activate already existing personal ideas, quests, and desires, a process I have referred to as cognitive recognition. A conversion does not imply that a cognitive framework is totally replaced by a new one. Rather, the exploration of something unknown, the study of a new belief and the convert's sense of change, take place within and are experienced through a personal world, through already salient ideas. Islam is appropriated and understood through a personal model that will be not only modified due to the integration of a new belief and experiences, but also concomitantly reinforced. Ayşe was captivated by the way Islam was practiced in her husband's home country, and the religious messages of social justice and communality appealed to her preexisting cognitive framework of social conscience and solidarity with the Third World. Integrating new ideas modifies her personal model but the religious ideas also confirm her already accepted important beliefs, providing her *personal model of social justice and solidarity with the Third World* with a religious dimension; Fatimah's emotion-laden dream and experience of *zikr* addressed her personal quest for spiritual meaning, and through a *personal model of spiritual yearning* she internalized Islam and Sufism. The salient models, intimately linked to the women's self-understanding, allows for the coexisting feeling of change and continuity.

Becoming Muslim reflects cognitive reconciliation, an intricate microprocess of meaning-making. Through cognitive reconciliation, constituting the conversion itself but also signifying identity formation, the self selectively chooses, refashions, and reconciles various experiences and public representations through the idiosyncratic world of thought-feelings, allowing for a sense of self-coherence despite change and appropriation of conflicting discourses. With cognitive reconciliation I refer to the psychological work of integrating personal meaning with, in this case, Muslim discourses, of bringing together old and new ideas and experiences, of synthesizing ongoing experiences. For example, Mariam reconciles childhood memories with spiritual experiences of Islamic and Sufi ideas. Through a *model of spiritual understanding and connection* she unites emotion-laden memories of dressing up as Mary and singing Bach's *St. Matthew Passion* with later ones doing fieldwork in a remote corner of the world as a young woman in her twenties and more recent experiences of herself as Muslim and a mother. Besides the parallel processes of change and continuity, the analysis of cognitive reconciliation identifies the indivisible and simultaneous processes of acquiring a religious belief (and culture in general) and identity formation.

The looping effects impelled by the conversion involve not only the making of personal versions of Islam, but also the women's sense of "Swedishness," "Americanness," and "femininity" that are reconsidered and given new meaning. Embracing a Muslim identity changes not only the convert's sense of being Swedish or American, but also her experience of femininity. The conversion to Islam has generated criticism of Western and Muslim ideals of female appearance, of materialism, and the lack of spirituality, moral values, and social integration, as well as praise for a culturally and religiously diverse society. Becoming Muslim has thus stimulated new ways of reflecting upon other belongings, triggering different alternative versions of Swedishness, Americanness, and femaleness.

We identify ourselves through available categories, but, as the converts' narratives so unmistakably lay bare, these categories have no inherent meaning but are rather infused by specific subjective experiences and charged emotional memories. The women's active engagement with the world spurs personal commentary on the categories employed by, but also imposed on, them. As we have seen, these looping effects of meaning become particularly palpable when Marianne, in an extended exploration of various face-to-face interactions and more anonymous encounters, employs categories and expressions such as "Swedish," "Muslim," "Muslim in a rather Swedish way," "immigrant," and "100 percent myself" to communicate different self-understandings. At first glance it is tempting to understand this as a shifting of multiple identities reflecting a fragmented self. My analysis of the account offers a different way of looking at it. In an intense and intricate sense-making, drawing on diverse classifications, Marianne attempts to communicate, to me as well as herself, her sense of self in relation to how others perceive her. Through the expression "100 percent Swedish" she communicates an experienced gap between selfhood and personhood when interacting with some Muslims, rather than a particular notion of national belonging. Further, when she feels "100 percent myself" her inner sense of self corresponds to the self reflected back to her in the interaction. Marianne renders here a state of mind in which she feels like she has been able to convey her personal meaning of being Muslim.

Due to prevalent stereotypes and a sometimes hostile environment, the convert's Muslim identity entails a heightened awareness of a targeted selfhood. In interaction with others (myself included) their identity as Muslim frequently becomes an object of critical examination as well as contemplation. The experiences of different interpersonal relationships such as Fatimah's mother's difficulties in accepting her daughter's religious identity, Ayşe's employer's discriminating and disparaging comments, and Marianne's family members' rejection of her Muslim identity trigger sensitivity in regard to one's sense of self.

One remedy for this tension is to tell one's story. The conversion itself compels the women to tell their stories of how they became Muslim. Just as the encounters the women described to me, the interview situation itself offered an opportunity to give me their personal conversion story, to try out different self-representations and resist socially imposed classifications. These compose social situations in which they can make sense of changes in life, not only for others but also for themselves. In the dialogue with me it becomes important, in response to what they believe others think about them, to counter an idea that they are somehow oppressed and to demonstrate their engagement in women's rights, issues of peace, social justice and solidarity, and their belief in God. As a critical part of the conversion itself, the conversion narrative serves as a salient means to reconcile and mediate between self and other, past and present, and inner and outer.

I wish to end the book with an excerpt from the last interview encounter I had with the women. It took place one day in June 2002 in Fatimah's kitchen. We talked about the things that had happened in our lives since the last time we first met and soon the topic of September 11 was brought up. Almost a whole year had passed since the terrorist attacks. It had shaken Fatimah just as everybody else; the event has made it harder to be a Muslim in some social interactions. One woman, for example, had stopped talking to her when she found out that Fatimah was Muslim. On the other hand, it had for the first time in fifteen years initiated a serious conversation with her mother about Islam, a conversation without defensive arguments.

And recently, I just spent a week with my parents, I was helping my mother, and she asked me one day the question of "what is your association of these terrorist guys overseas?" But unlike other times when she has asked me questions, this was a big charged question . . . But this was a question that was really easy for me to answer, because from my standpoint I have *no* association with these people [laughs]. So, you know, for me it is as hard as it is for you. I don't understand it, it is scary to me. Evil. That was actually the first time when we ended up having a conversation about Islam. Otherwise we don't talk about it very much. But it is not as charged as it used to be. It is not as scary to them as it used to be.

Anna: It has been a while.

Fatimah: Yeah, it has been fifteen years and some . . . close to that, I think. They know my family, they love them. And they know that they are not weird terrorist people. It is not as charged as it used to be.

Anna: I can imagine that there is a gap between what they see on TV and then their daughter's life that does not fit into that.

Fatimah: Yeah, I think if I was another kind of personality and had taken . . . and also, and this is where I think Sufism is so important, because both the sheikhs preach so much about tolerance and love and that Islam is peace. You know, it is inner peace, it is outer peace . . . You know I have held on very strongly to be an American. And I don't try to pretend that I'm from another country or anything. No, because that is one thing I feel about being American, that I can be Muslim in America. Even if I have to admit that it has gotten harder. For me Islam is very personal. I don't identify as strongly with the Muslims of the world, I didn't convert as a part of a social network. I converted because of the prayers and the connection to Allah. For very personal reasons.

The categories of "Muslim" and "Islam" have become even more charged and for Fatimah it is important to point to the "very personal reasons" behind her conversion. In discussion with her mother, Fatimah has to act upon the categories evoked in the question and make clear her position. Then in the conversation with me about this interaction, meanings about being a Muslim are triggered and verbalized again. In dialogue with the world, Fatimah feels an urgency to mediate the personal meaning assigned to Islam, that of inner peace and spiritual connection and how her Muslim and American identities relate to each other. These engagements reveal the women's serious attempts to bring an inner sense of self into some kind of equilibrium with who she is perceived to be by others. The women, and all of us for that matter, want to be understood within their own biographical particularities.

Notes

Chapter 1 Conversion and Identity Formation

1. Sometimes "reverts" is used instead of "converts" (see also note 2). The word "reverts" stresses the Muslim idea that they are coming back to what they, and all of us, were born as. I will, however, use the more common concept of convert here, implying, in general terms, a person who has left a worldview for another.
2. There are several Internet sites with conversion stories of "reverts" to Islam, see, for example, http: //thetruereligion.org, http: //www.islamfortoday. com/converts.htm, and http: //www.convertstoislam.com.
3. Similarly to the notion of culture, I do not understand a belief system to be a coherent and consistent whole. Tanya Luhrmann argues that a belief or belief system cannot be treated as a consistent set of propositions. "[I]t is optimistic to think that people have an ordered set of beliefs about a particular endeavour which forms a consistent set with other beliefs which together describe the totality of thought and action" (Luhrmann 1989:309).
4. A souq, the Arabic word for market, is a place where one can buy clothes, fabric, food, spices, and handcrafts.
5. The celebration of the day of Lucia is an old tradition, originating from the celebration of the Sicilian Saint Lucia. It is celebrated on December 13 in Scandinavia in the early hours of the morning with a procession of singing women, men, and children led by Lucia herself, a woman dressed in white, wearing a crown of candles. Besides Advent, Lucia Day symbolically begins the Christmas celebrations, bringing hope and light to the darkest month in the year.
6. Larry Poston (1992) shows how Islam's missionary activity had to change in its method of outreach in the contemporary West. While early expansion of Islam relied upon an "external-institutional" missionary approach, Muslims in the West have, as a response to the new demands of Western social contexts, adopted a, what he calls, "internal-personal" approach that instead focuses on the conversion of individuals and aims to influence society from the bottom upwards.

7. See also work on female converts to Islam in the United Kingdom (Ball 1987) as well as in Germany (Hofmann 1997) and conversion literature of a biographical character (Lang 1994).

8. Daniel Linger distinguishes between "public worlds" and "personal worlds" as two separate, but not closed, systems that interact with each other. Criticizing "post-Durkheimian Anthropology" and its discursivist perspective, he argues that in order to understand human worlds we need to study both (Linger 2005:12f.).

9. As Martin Sökefeld stresses, to avoid the danger of voluntarism, agency is best understood in close connection to the reflective self, which recognizes the conditions for action, "the limits of agency." Hence, self, reflexivity, and agency are understood as inseparable (1999:430).

10. Roy D'Andrade, Claudia Strauss, and Naomi Quinn are some of the anthropologists who have developed similar approaches, linking mind to culture by analyzing mental models and how experience and cultural messages are internalized, and whose work I have been inspired by. D'Andrade has pointed out that the most fruitful place to start the study of culture is not the collection of public messages, since messages and meanings are not the same, but individual "meaning systems," which are mental structures and processes. He states, "Meanings represent the world, create cultural entities, direct one to do certain things, and evoke certain feelings" (D'Andrade 1984:96). Hence, cultural meanings involve psychological processes. This theoretical point of departure challenges a general presupposition in much symbolist, structuralist, and interpretative anthropology that symbols, language, or discourses determine how we experience the world (D'Andrade 1995:148). With attention to experience, emotion, and agency as well as how people make sense of reality according to their own categories, and not those of the anthropologist, approaches of psychological anthropology have several standpoints in common with the tradition of phenomenology (see e.g., Jackson 1996).

11. With emphasis on thought-feelings, a phrase by Unni Wikan (1989), the approach wants to avoid the Western philosophical dichotomy of thought and emotion. There is a tendency within anthropology to dismiss the affective aspect of culture based on the idea that reason and feeling contradict each other, an idea that seems to be embedded in American and European culture (D'Andrade 1984:101).

12. The scope and approaches of cognitive anthropology are expansive. See D'Andrade 1995 for a thorough account and discussion of the development of the tradition of cognitive anthropology. The main problem of cognitive anthropology is how cultural knowledge is organized in the mind. The early cognitive anthropology in the 1950s to early 1970s was much influenced by the semantic analysis of terminology classifications and taxonomies, but a methodological and theoretical shift in the mid-1970s contributed to an increased focus on psychological processes. The move from the linguistic emphasis and the view of culture as

knowledge to more psychologically oriented questions regarding the relationship between mental and public representations and how culture is acquired, understood, and communicated implied that the previous strong dependence of thought upon language was broken. With the increased interest in schema theory and cultural models in the 1980s, researchers have tried ways of further exploring the psychological aspects of culture, not only how culture is organized in the mind but also how it is linked to emotion and motivation (D'Andrade 1995).

13. See also Daniel Linger's *Dangerous Encounters: Meanings of Violence in a Brazilian City* (1992).

14. By "internalized" I do not suggest that a religious system or representation is "taken in" as a whole (cf. Strauss and Quinn 1997:9). On the contrary, this study demonstrates how the conversion, as an ongoing process of reconciling old and new, does not totally transform the self but rather how the religious belief is appropriated through a particular personal model.

15. Written from a linguistic perspective, another quite different anthropological approach to conversion is offered by Susan Harding (1987, 2000) in her interesting account of Jerry Falwell's church and born-again Christians, in which she develops a theory of the vital role of language as a medium through which selves are formed and reformed. Conversion is here understood as a process of acquiring a specific language, when a listener becomes a speaker. It is the Bible-based language that converts the listener's mind and, later, identity. Harding asserts, "speaking is believing" (2000:60).

16. She describes "interpretative drift" as a "slow, often unacknowledged shift in someone's manner of interpreting events as they become involved with a particular activity" (1989:312).

17. Focusing on the performance of the conversion narrative, Peter Stromberg shows that the ideological language of Evangelical Christianity is not just a means of narrating what happened but also a continuing means of communicating and resolving emotional problems of meaning and conflicting desires. Rather than a "one-time transformation" in which the conflicts supposedly should disappear, the conversion is an ongoing process of resolution. The conflicts are not resolved, but persist in the very moment of narration, and through the ideological language the convert can negotiate and come to terms with his or her emotional ambivalence (1990). In another article (1985), he analyzes, starting from a description of St. Augustine's conversion, what he calls an "impression point," a point where the new religion is internalized and becomes conceivable and meaningful in relation to the particular problems and quests of the individual.

18. I have here in mind studies that have become "classics" within the literature of conversion, such as William James's *The Varieties of Religious Experience* (1903), Edwin Starbuck's *The Psychology of Religion* (1901), and John Lofland and Rodney Stark's work on a millenarian

cult on the West Coast of the United States (1965). Common to these, and other more recent sociological and psychological studies, is the mapping of certain factors and external psychological and social conditions underlying the conversion as well as different stages. In general, the purpose of these studies seems to be to design a pattern of conversion, a model of the typical convert, involving age, background, and reasons, applicable to any other study of conversion. They are also often, as pointed out by Larry Poston, conducted from a Christian perspective and in a Christian environment. Reflecting a traditional Western paradigm, even the supposedly "neutralist" or "secularist" approaches are often characterized by Western and Christian assumptions (Poston 1992:154). Furthermore, the "neutralist" or "secularist" assumption is problematic. Lewis Rambo expresses strong criticism toward the "reductionism" in much of the studies on religion and conversion in the humanities and social sciences. He writes, "The secular assumptions that pervade the human sciences result in an often derogatory tone by those purporting to study religious phenomena" (1993:xiv).

19. Even if it draws on a quite different theoretical framework, the work of Anne Bolin (1994) also emphasizes the aspects of process and ongoing acquisition of new identity in her study on male-to-female transsexuals. By referring to the phenomenon of "rite of passage" she discusses how the process of "becoming" is characterized by a transformation of personal identity, social identity, and physiology.

20. This implies avoiding theoretical hypotheses distant, and without reference, to the experiences expressed by those being studied.

21. I interviewed them twice with the exception of Zarah who did not want to be interviewed a second time and Marianne who was interviewed three times.

22. The interviews were rather unstructured in the sense that I had only some major themes to cover but let the women otherwise elaborate freely around the issues that appeared to be of most importance to them. The women were interviewed in their homes and in some cases at their work, at a Muslim organization or, as on one occasion, in a café. Before the tape recorder was switched on (none of the women objected to this) I told them again about my purpose and the study as a whole. I then asked them to tell me about their upbringing, when and where they were born, and then later about their first encounter with Islam and Muslims. Other "themes" concerned changes in lifestyle, ideas that were still important to them and different encounters and relationships with parents and friends. Central questions of mine such as the meaning of Islam were not put forward explicitly but were instead elaborated and developed by the women themselves throughout the interviews. Similar questions are often difficult to answer when posed directly.

23. The *hadith* literature comprises the explanation of the Qur'an. Referring to the Prophet's example, the *hadith* specify the guidelines offered in the Qur'an. For a thorough discussion see Roald 2001.

24. http://www.journeytoislam.com/become_muslim/how_become_muslim.htm (October 20, 2004).

25. This is said with an emphasis on "rather." The idea of Sweden as being traditionally ethnic homogenous until very recently has been rightly criticized (Gaunt 1996). There have also been several different factions within the "Swedish religion." For example, nonconformist movements have often challenged the Swedish church.

26. These numbers were mentioned by the islamologist Christer Hedin in one of the leading Swedish newspaper, *Dagens Nyheter*, January 20, 2002.

27. The development and presence of Islam in a European context has been explored in, to mention some, Jørgen Nielsen's *Muslims in Western Europe* (1992) and *Toward a European Islam* (1999), Anne Sophie Roald's *Women in Islam: The Western Experience* (2001), and Wasif Shadid and Sjoerd van Koningsveld's *Religious Freedom and the Position of Islam in Western Europe: Opportunities and Obstacles in the Acquisition of Equal Rights* (1994). Yvonne Haddad, another prominent researcher in the area, has written several studies about Muslims and Islamic values and thoughts in the United States (see e.g., 1987, 1991). *Muslims on the Americanization Path?* (Haddad and Esposito 1998) is another volume of articles about the variety of Muslim experiences in America covering significant themes like the tension between tradition and change, the relation of law to society and the role of religion in cultural and national identity.

28. I owe this discussion to Daniel Linger (personal communication). His work on Japanese Brazilians in Japan deals with similar questions regarding national categories (2001, 2003).

Chapter 2   A Step toward the Unknown

1. Sufism is a mystical dimension of Islam and there are today many different Sufi orders over the world. It is often depicted as an open, inclusive, and tolerant "movement" within Islam. With its focus on spirituality and mysticism it has attracted people in the West. It is also characterized by the idea that every Sufi should follow a Sufi sheikh, a spiritual leader. For an introduction to the development of Sufism in the United States see, for example, Kinney (1994).

2. This is the official Islamic ceremony of conversion that involves saying the declaration of faith—"There is no God but Allah and Mohammad is His messenger"—in front of an imam and a witness.

Chapter 3   The Conversion Narrative

1. Unni Wikan has also ventilated criticism against anthropologists' heavy reliance on the medium of the spoken language while overlooking the

acts through which people fashion themselves and particularly the crucial aspect of silence as "lived predicament," that is, what is not said. She asserts that silence also "affects our conception of life and ourselves" (1995:265).

2. James Peacock and Dorothy Holland (1993) seem to advocate a similar approach that they call "processual," an approach that places emphasis on the telling of life stories as an important event in psychological and sociocultural life.

3. These "coherence systems," or as Charlotte Linde calls them elsewhere, "explanatory systems" (1987), are systems of beliefs at an intermediate level between "*common sense*, the beliefs and relations of beliefs that any person in the culture may be assumed to know, if not to share, and *expert systems*, which are beliefs and relations among beliefs held, understood, and used by experts in a particular domain" (1987:343). By the use of explanatory systems, that is, popular versions of expert systems, the speaker does not only construct coherence in the story but also provides an evaluation of what the story means.

4. In a *zikr*, sometimes spelled *dhikr*, the names of Allah are repeated to reach a spiritual state of "remembrance of God," the very meaning of the word.

Chapter 4   Personal Models of Spirituality and Social Conscience

1. Zarah, who was only interviewed once, is not discussed in these chapters. See discussion in chapter 3.

2. See Bennetta Jules-Rosette (1976), who describes her own conversion experience during research on an indigenous African church, the Apostles of John Maranke.

3. See note 1 in chapter 2.

4. Home schooling seems to be a rather common phenomenon in the area where she and her family live, not only among Muslims but also in born-again Christian families. Fatimah also had her son and daughter in home school. The women themselves help their children with the readings assigned by a supervised teacher who comes regularly to check that the children do the assigned homework and learn what they are supposed to do.

5. Martin Lings was a prominent representative of the movement of Shadhili-Alawi tradition. He wrote *What is Sufism?* and *A Sufi Saint of the Twentieth Century: Shaikh Ahmad Al-'Alawî.*

6. This is partially the same quote as on page 74 in chapter 3.

7. Since Fatimah is not her real name for reasons of anonymity (but the name she chose herself for the study), the Arabic meaning of Fatimah does not refer to Mary.

Chapter 5   Personal Models of Gender

1. We should be careful not to fall into generalizations when talking about "American," "Muslim," or "Swedish" models of gender relations. There are of course many versions and practices of them, which is also my purpose to show, or, as Mariam pointed out, "I do not think there is a general American [female] role." Still the models do, I believe, represent rather widespread discursive ideas, to which the women have to relate to in one way or another. The "Swedish" and "American" versions are described as rather similar and could be referred to, and again in general, as a Western representation of the female role and obligation.

2. The last sentences of this quotation compose the same account as in the very beginning of chapter 1.

3. Katherine Ewing (1990) adopts a psychoanalytic approach to explain how the psyche organizes and internalizes interpersonal experiences. Drawing on psychoanalytical terms such as "transference" and "condensation" she explains how the individual constructs an illusionary sense of wholeness and continuity out of what are inconsistent experiences.

Chapter 6   The Veil and Alternative Femininities

1. I am referring here to the common Islamic dress code of covering everything but face and hands (there are of course exceptions in the clothing, depending on the context as well as on the interpretation of the *suras* and the *hadiths*). Among the women I have met, with the exception of Fatimah and Mariam, the most common clothing is a large headscarf that is fastened below the chin covering hair, ears, neck, shoulders, and chest, and a long, loose dress or a long-sleeved shirt with pants.

Chapter 7   Looping Effects of Meaning

1. Marianne's confrontation with her own reflection is similar to Dorinne Kondo's account of how she during fieldwork in Japan got a glimpse of herself in a shop's display case. Both accounts deal with an inner reflection of self and a sense of becoming "the Other," but Kondo's analysis differs from mine since she elaborates on a theory of the fragmentation of self. However, Kondo also touches upon a profound experience of the otherness of one's self: ". . . I noticed someone who looked terribly familiar: a typical young housewife, clad in slip-on sandals and the loose, cotton shift called 'home wear' (*hōmu wea*), a woman walking with a characteristically Japanese bend to the knees and a sliding of the feet.

Suddenly I clutched the handle of the stroller to steady myself as a wave of dizziness washed over me, for I realized I had caught a glimpse of nothing less than my own reflection" (Kondo 1990:16f.).
2. Almsgiving, in Arabic *zakat*, is one of the five creeds.
3. Daniel Linger finds a similar sense of belonging negotiated among Japanese Brazilians when visiting Brazilian restaurants (Linger 2001:92).
4. African Americans, next to Arabs and South Asians, are the largest groups representing Islam in the United States (Schmidt 2004). For a discussion on the historical background and the ideological conflicts within the African American community and its relation to multi-ethnic Muslim immigrant groups, see Nuruddin (1998) and Dannin (1998), respectively. Both of the authors point to the tension between racial identity and Islam within African American movements and local communities.

Chapter 8   Family, Work, Sisterhood

1. *Sydsvenska Dagbladet*, August 13, 1999.
2. *The Economist*, August 28, 1999. It is important to note that it is not only in Western countries that there has been opposition to veiling but in Muslim countries as well. For example, in Java, Indonesia, the state tried to ban veiling in the public schools. This is still in a country quite influenced by Islam where about ninety percent of the population is Muslim (Brenner 1996:676).
3. *Sydsvenska Dagbladet*, August 2, 1998.
4. Ibid., August 21, 1998.
5. Since May 1999, there has been a law regarding ethnic discrimination besides the act on freedom of religion from 1951.
6. An international comparison from 1997 showed that Sweden, together with Norway and Denmark, is one of few industrial countries where the number of unemployed among foreign citizens is at least 2.5 times as high as among the countries' own citizens. *Dagens Nyheter*, April 20, 1997.
7. Interestingly, I have encountered similar resistance to the idea that Muslim women can embrace feminist agendas among students in Women's Studies classes.

References

Ahmed, Leila. 1992. *Women and Gender in Islam: Historical Roots of a Modern Debate*. New Haven and London: Yale University Press.

Alloula, Malek. 1986. *The Colonial Harem*. Minneapolis: University of Minnesota Press.

Alsmark, Gunnar. 2001. Masooma—a Ugandan-Asian Muslim Swede. In *Beyond Integration: Challenges of Belonging in Diaspora and Exile*, ed. M. Povrzanović Frykman, pp. 85–100. Lund: Nordic Academic Press.

Anway, Carol L. 1996. *Daughters of another Path: Experiences of American Women Choosing Islam*. Lee's Summit, MO: Yawna Publications.

———. 1998. American Women Choosing Islam. In *Muslims on the Americanization Path?*, ed. Y. Y. Haddad and J. L. Esposito, pp. 179–98. Atlanta, GA: Scholars Press.

Ball, Harfiyah. 1987. *Islamic Life: Why British Women Embrace Islam?* In Muslim Community Survey series. Leicester, U.K.: Muslim Youth Education Council.

Berg, Magnus. 2001. Orienten på en höft. Orientalisk dans i Sverige. In *Där hemma, här borta: Möten med Orienten i Sverige och Norge*, ed. Å. Andersson, M. Berg, and S. Natland. Stockholm: Carlssons Bokförlag.

Berger, Peter L., and Thomas Luckmann. 1984 (1966). *The Social Construction of Reality: A Treatise in the Sociology of Knowledge*. Harmondsworth: Penguin Books.

Billiet, Richard. 1979. *Conversion to Islam in the Medieval Period*. Cambridge: Harvard University Press.

Bloch, Maurice. 1998. *How We Think They Think: Anthropological Approaches to Cognition, Memory, and Literacy*. Boulder: Westview Press.

Bolin, Anne. 1994. Transcending and Transgendering: Male-to-Female Transsexuals, Dichotomy and Diversity. In *Third Sex, Third Gender: Beyond Sexual Dimorphism in Culture and History*, ed. G. Herdt, pp. 447–85. New York: Zone Books.

Brenner, Suzanne. 1996. Reconstructing Self and Society: Javanese Muslim Women and "the Veil." *American Ethnologist* 23(4): 673–97.

Buckser, Andrew, and Stephen D. Glazier. 2003. *The Anthropology of Religious Conversion*. Lanham: Rowman & Littlefield Publishers.

Bulbeck, Chilla. 1998. *Re-Orienting Western Feminisms: Women's Diversity in a Postcolonial World*. Cambridge: Cambridge University Press.

Chodorow, Nancy J. 1999. *The Power of Feelings: Personal Meaning in Psychoanalysis, Gender, and Culture.* New Haven and London: Yale University Press.

Cohen, Anthony P. 1994. *Self Consciousness: An Alternative Anthropology of Identity.* London and New York: Routledge.

Collins, Peter. 1998. Negotiating Selves: Reflections on "Unstructured" Interviewing. *Sociological Research Online* 3(3). (Internet document http://www.socresonline.org.uk/socresonline/ 3/3/2.html.)

Corbin, Henry. 1966. The Visionary Dream in Islamic Spirituality. In *The Dream in Human Societies,* ed. G. von Grunebaum and R. Caillois, pp. 381–408. Los Angeles: University of California Press.

Crapanzano, Vincent. 1980. *Tuhami: Portrait of a Moroccan.* Chicago: Chicago University Press.

D'Andrade, Roy G. 1984. Cultural Meaning Systems. In *Culture Theory: Essays on Mind, Self, and Emotion,* ed. R. A. Schweder and R. LeVine, pp. 88–119. Cambridge: Cambridge University Press.

———. 1987. A Folk Model of the Mind. In *Cultural Models of Language and Thought,* ed. D. Holland and N. Quinn, pp. 112–48. Cambridge: Cambridge University Press.

———. 1995. *The Development of Cognitive Anthropology.* Cambridge: Cambridge University Press.

D'Andrade, Roy G., and Claudia Strauss, eds. 1992. *Human Motives and Cultural Models.* Cambridge: Cambridge University Press.

Dannin, Robert. 1998. Understanding the Multi-Ethnic Dilemma of African-American Muslims. In *Muslims on the Americanization Path?,* ed. Y. Y. Haddad and J. L. Esposito, pp. 331–58. Atlanta, GA: Scholars Press.

Douglas, Mary. 1966. *Purity and Danger: An Analysis of the Concepts of Pollution and Taboo.* New York: Praeger.

Dutton, Yasin. 1999. Conversion to Islam: The Qur'anic Paradigm. In *Religious Conversion. Contemporary Practices and Controversies,* ed. C. Lamb and M. D. Bryant, pp. 151–65. London and New York: Cassell.

Edgar, Iain. 1994. Dream Imagery Becomes Social Experience: The Cultural Elucidation of Dream Interpretation. In *Anthropology and Psychoanalysis: An Encounter through Culture,* ed. S. Heald and A. Deluz, pp. 99–113. London and New York: Routledge.

El Guindi, Fadwa. 1999. *Veil: Modesty, Privacy, and Resistance.* Oxford and New York: Berg.

El-Solh, Camilla Fawzi, and Judy Mabro. 1994. *Muslim Women's Choices: Religious Belief and Social Reality.* Providence: Berg.

Eriksen, Thomas H. 1994. *Kulturterrorismen: Et uppgjør med tanken om kulturell renhet.* Oslo: Spartacus Forlag.

Esposito, John L. 1998. Muslims in America or American Muslims. In *Muslims on the Americanization Path?,* ed. Y. Y. Haddad and J. L. Esposito, pp. 3–17. Atlanta, GA: Scholars Press.

Ewing, Katherine Pratt. 1990. The Illusion of Wholeness: Culture, Self, and the Experience of Inconsistency. *Ethos* 18(3): 251–78.

———. 1997. *Arguing Sainthood: Modernity, Psychoanalysis, and Islam.* Durham and London: Duke University Press.

———. 1998. Crossing Borders and Transgressing Boundaries: Metaphors for Negotiating Multiple Identities. *Ethos* 26(2): 262–67.

Gaunt, David. 1996. Ethnic relations in the Swedish empire 1000–1800. In *To Make the World Safe for Diversity*, ed. Å. Daun. Botkyrka: The Swedish Immigrant Institute.

Gerholm, Tomas. 1988. Three European Intellectuals as Converts to Islam: Cultural Mediators or Social Critics? In *The New Islamic Presence in Western Europe*, ed. T. Gerholm and Y. G. Lithman, pp. 263–77. London and New York: Mansell Publishing Limited.

Goffman, Ervin. 1959. *The Presentation of the Self in Everyday Life.* Garden City, NY: Anchor Books.

Gregg, Gary S. 1998. Culture, Personality, and the Multiplicity of Identity: Evidence from North African Life Narratives. *Ethos* 26(2): 120–52.

Hacking, Ian. 1995. *Rewriting the Soul. Multiple Personalities and the Sciences of Memory.* Princeton: Princeton University Press.

———. 1999. *The Social Construction of What?* Cambridge and London: Harvard University Press.

Haddad, Yvonne Yazbeck. 1987. *Islamic Values in the United States: A Comparative Study.* New York: Oxford University Press.

———. 1991. *The Muslims of America.* New York: Oxford University Press.

———. 1998. Islam and Gender: Dilemmas in the Changing Arab World. In *Islam, Gender, and Social Change*, ed. Y.Y Haddad and J. L. Esposito, pp. 3–29. New York: Oxford University Press.

Haddad, Yvonne Yazbeck, and John L. Esposito, eds. 1998. *Muslims on the Americanization Path?* Atlanta, Georgia: Scholars Press.

Harding, Susan. 1987. Convicted by the Holy Spirit: The Rhetoric of Fundamental Baptist Conversion. *American Ethnology* 14(1): 167–81.

———. 2000. *The Book of Jerry Falwell: Fundamentalist Language and Politics.* Princeton: Princeton University Press.

Hastrup, Kirsten. 1992. Writing Ethnography: State of the Art. In *Anthropology and Autobiography*, ed. J. Okely and H. Callaway, pp. 116–33. London and New York: Routledge.

Hofmann, Gabriele. 1997. *Muslimin werden: Frauen in Deutschland konvertieren zum Islam.* Frankfurt and Main: Inst. für Kulturanthropologie und Europäische Ethnologie.

Hollan, Douglas. 1997. The Relevance of Person-Centered Ethnography to Cross-Cultural Psychiatry. *Transcultural Psychiatry* 34(2): 219–34.

———. 2000. Constructivist Models of Mind, Contemporary Psychoanalysis, and the Development of Culture Theory. *American Anthropologist* 102(3): 538–55.

———. 2001. Dreaming in a Global World. Paper presented at the Biennial SPA meeting in Decatur, GA, 2001.

Holland, Dorothy. 1992. How Cultural Systems Become Desire: A Case Study of American Romance. In *Human Motives and Cultural Models*, ed. R. G. D'Andrade and C. Strauss, pp. 61–89. Cambridge: Cambridge University Press.

Holland, Dorothy, and Debra Skinner. 1987. Prestige and Intimacy: The Cultural Models behind Americans' Talk about Gender Types. In *Cultural Models in Language and Thought*, ed. D. Holland and N. Quinn, pp. 78–111. Cambridge: Cambridge University Press.

Holland, Dorothy, and Naomi Quinn, eds. 1987. *Cultural Models in Language and Thought*. Cambridge: Cambridge University Press.

Holland, Dorothy, William Lachicotte Jr., Debra Skinner, and Carole Cain, eds. 1998. *Identity and Agency in the Cultural Worlds*. Cambridge and London: Harvard University Press.

Hoodfar, Homa. 1991. Return to the Veil: Personal Strategy and Public Participation in Egypt. In *Working Women: International Perspectives on Labour and Gender Ideology*, ed. N. Redclift and M. T. Sinclair, pp. 104–24. London and New York: Routledge.

Huntington, Samuel. 1993. The Clash of Civilizations. *Foreign Affairs* Summer 72(3): 22–49.

Jackson, Michael. 1996. Introduction. In *Things as they Are: New Directions in Phenomenological Anthropology*, ed. M. Jackson, pp. 1–50. Bloomington and Indianapolis: Indiana University Press.

Jameson, Frederic. 1991. *Postmodernism, or, the Cultural Logic of Late Capitalism*. Durham, NC: Duke University Press.

Jules-Rosette, Bennetta. 1976. The Conversion Experience: The Apostle of John Maranke. *Journal of Religion in Africa* Vol. VII: 132–64.

Kinney, Jay. 1994. Sufism Comes to America. *Gnosis Magazine* Winter 1994(30): 18–23.

Kondo, Dorinne K. 1990. *Crafting Selves: Power, Gender, and Discourses of Identity in a Japanese Workplace*. Chicago: The University of Chicago Press.

Köse, Ali. 1996. *Conversion to Islam: A Study of Native British Converts*. London: Kegan Paul International.

Lang, Jeffrey. 1994. *Struggling to Surrender: Some Impressions of an American Convert to Islam*. Beltsville: Amana Publications.

Levtzion, Nehemia. 1979. *Conversion to Islam*. New York: Holmes and Meier Publishers.

Levy, Robert I. 1994. Person-Centered Anthropology. In *Assessing Cultural Anthropology*, ed. R. Borofsky, pp. 180–89. New York: McGraw-Hill.

Linde, Charlotte. 1986. Private Stories in Public Discourse: Narrative Analysis in the Social Sciences. *Poetics* 15(1–2): 183–202.

———. 1987. Explanatory Systems in Oral Life Stories. In *Cultural Models in Language and Thought*, ed. D. Holland and N. Quinn, pp. 343–67. Cambridge: Cambridge University Press.

———. 1993. *Life Stories: The Creation of Coherence*. New York: Oxford University Press.

Linger, Daniel T. 1992. *Dangerous Encounters: Meanings of Violence in a Brazilian City*. Stanford: Stanford University Press.

———. 2001. *No One Home: Brazilian Selves Remade in Japan*. Stanford: Stanford University Press.

———. 2003. Do Japanese Brazilians Exist? In *Searching for Home Abroad: Japanese Brazilians and Transnationalism*, ed. J. Lesser, pp. 201–14. Durham: Duke University Press.

———. 2005. *Anthropology through a Double Lens: Public and Personal Worlds in Human Theory*. Philadelphia: University of Pennsylvania Press.

Löfgren, Orvar. 1993. Materializing the Nation in Sweden and America. *Ethos* 1993(3–4): 161–96.

Lofland, John, and Rodney Stark. 1965. Becoming a World-Saver: A Theory of Conversion to a Deviant Perspective. *American Sociological Review* 30(6): 862–75.

Luhrmann, Tanya M. 1989. *Persuations of the Witch's Craft: Ritual Magic in Contemporary England*. Cambridge: Harvard University Press.

Macleod, Arlene E. 1991. *Accommodating Protest: Working Women, the New Veiling, and Change in Cairo*. New York: Columbia University Press.

Mahoney, Maureen A., and Barbara Yngvesson. 1992. The Construction of Subjectivity and the Paradox of Resistance: Reintegrating Feminist Anthropology and Psychology. *Signs* 18(1): 44–73.

Malkki, Lisa. 1992. National Geographic: The Rooting of Peoples and the Territorialization of National Identity among Scholars and Refugees. *Cultural Anthropology* 7(1): 24–44.

Mansson, Anna. 2000. Möten mellan "svenskt" och "muslimskt." In *Att möta främlingar*, ed. G. Rystad and S. Lundberg, pp. 259–88. Lund: Arkiv Förlag.

Mernissi, Fatima. 1975. *Beyond the Veil: Male-Female Dynamics in Modern Muslim Society*. London: Al Saqi Books.

———. 1991. *The Woman in Islam: An Historical and Theological Inquiry*. Oxford: Blackwell.

———. 2001. *Scheherazade Goes West: Different Cultures, Different Harems*. New York: Washington Square Press.

Mohanty, Chandra. 2003. *Feminism without Borders: Decolonizing Theory, Practicing solidarity*. Durham: Duke University Press.

Moore, Kathleen. 1998. The *Hijab* and Religious Liberty: Anti-Discrimination Law and Muslim Women in the United States. In *Muslims on the Americanization Path?*, ed. Y. Y. Haddad and J. L. Esposito, pp. 129–58. Atlanta, GA: Scholars Press.

Narayan, Uma. 1997. *Dislocating Cultures. Identities, Traditions, and Third World Feminism*. New York and London: Routledge.

Nielsen, Jørgen. 1992. *Muslims in Western Europe*. Edinburgh: Edinburgh University Press.

———. 1999. *Toward a European Islam*. London: Macmillan.

Nuruddin, Yusuf. 1998. African-American Muslims and the Question of Identity: Between Traditional Islam, African Heritage, and the American Way. In *Muslims on the Americanization Path?*, ed. Y. Y. Haddad and J. L. Esposito, pp. 267–330. Atlanta, GA: Scholars Press.

Obeyesekere, Gananath. 1981. *Medusa's Hair: A Study in Personal and Cultural Symbols.* Chicago: University of Chicago Press.

Otterbeck, Jonas. 2000. *Islam på svenska: Tidskriften Salaam och islams globalisering.* Lund Studies in History of Religions Vol. 11.

Peacock, James L., and Dorothy C. Holland. 1993. The Narrated Self: Life Stories in Process. *Ethos* 21(4): 367–83.

Poston. Larry. 1992. *Islamic Da'wah in the West: Muslim Missionary Activity and the Dynamics of Conversion to Islam.* New York: Oxford University Press.

Pred, Allan. 1998. Memory and the Cultural Reworking of Crisis: Racisms and the Current Moment of Danger in Sweden, Or Wanting It Like Before. *Society and Space* 16(6): 635–64.

Quinn, Naomi. 1992. The Motivational Force of Self-Understanding: Evidence from Wives' Inner Conflicts. In *Human Motives and Cultural Models*, ed. R. G. D'Andrade and C. Strauss, pp. 90–126. Cambridge: Cambridge University Press.

Quinn, Naomi, and Dorothy Holland. 1987. Culture and Cognition. In *Cultural Models in Language and Thought*, ed. D. Holland and N. Quinn, pp. 3–40. Cambridge: Cambridge University Press.

Rambo, Lewis R. 1993. *Understanding Religious Conversion.* New Haven and London: Yale University Press.

———. 2003. Anthropology and the Study of Conversion. In *The Anthropology of Religious Conversion*, ed. A. Buckser and S. D. Glazier, pp. 211–22. Lanham: Rowman & Littlefield Publishers.

Rapport, Nigel, and Andrew Dawson. 1998. The Topic and the Book. In *Migrants of Identity: Perceptions of Home in a World of Movement*, ed. N. Rapport and A. Dawson, pp. 3–17. Oxford: Berg.

Roald, Anne Sophie. 2001. *Women in Islam: The Western Experience.* London and New York: Routledge.

Sachs Norris, Rebecca. 2003. Converting to What? Embodied Culture and the Adoption of New Beliefs In *The Anthropology of Religious Conversion*, ed. A. Buckser and S. D. Glazier, pp. 171–81. Lanham: Rowman & Littlefield Publishers.

Said, Edward W. 1978. *Orientalism.* New York: Pantheon Books.

———. 1994. *Culture and Imperialism.* New York: Vintage Books.

———. 1997. *Covering Islam. How the Media and the Experts Determine How We See the Rest of the World.* New York: Vintage Books.

Sander, Åke. 1991. The Road from Musalla to Mosque: The Process of Integration and Institutionalization of Islam in Sweden. In *The Integration of Islam and Hinduism in Western Europe*, ed. W. A. R. Shadid and P. S. van Koningsveld, pp. 62–88. Kampen: Kok Pharos.

———. 1995. Rasismens varp och trasor. In *Rasismens varp och trasor: En antologi om främlingsfientlighet och rasism*, pp. 132–66. Norrköping: Statens invandrarverk.

———. 1997. To What Extent is the Swedish Muslim Religious? In *Islam in Europe: The Politics of Religion and Community*, ed. S. Vertovec and C. Peach, pp. 179–210. Basingstoke: Macmillan.

Sapir, Edward. 1949 (1928). The Meaning of Religion. In *Selected Writings in Language, Culture, and Personality*, ed. D. G. Mandelbaum, pp. 346–56. Berkeley and Los Angeles: University of California Press.

Schmidt, Garbi. 2004. *Islam in Urban America: Sunni Muslims in Chicago*. Philadelphia: Temple University Press.

Schwartz, Theodore. 1978. Where is the Culture? Personality as the Distributive Locus of Culture. In *The Making of Psychological Anthropology*, ed. G. D. Spindler, pp. 419–41. Berkeley: University of California Press.

Shadid, Wasif A. R., and P. Sjoerd von Koningsveld. 1994. *Religious Freedom and the Position of Islam in Western Europe: Opportunities and Obstacles in the Acquisition of Equal Rights*. Kampen: Kok Pharos.

Skinner, Debra, and Dorothy Holland. 1998. Contested Selves, Contested Femininities: Selves and Society in Process. In *Selves in Time and Place: Identities, Experience, and History in Nepal*, ed. D. Skinner, A. Pach III, and D. Holland, pp. 87–110. Lanham: Rowman & Littlefield Publishers.

Sökefeld, Martin. 1999. Debating Self, Identity, and Culture in Anthropology. *Current Anthropology* 40(4): 417–47.

Sperber, Dan. 1996. *Explaining Culture: A Naturalistic Approach*. Oxford: Blackwell.

Spiro, Melford. 1987 (1982). Collective Representations and Mental Representations in Religious Symbol Systems. In *Culture and Human Nature: Theoretical Papers of Melford E. Spiro*, ed. B. Kilborne and L.L. Langness, pp. 161–84. Chicago: Chicago of University Press.

Starbuck, Edwin. 1901. *The Psychology of Religion: An Empirical Study of the Growth of Religious Consciousness*. New York: Scribner.

Strauss, Claudia. 1992a. Models and Motives. In *Human Motives and Cultural Models*, ed. R. G. D'Andrade and C. Strauss, pp. 1–20. Cambridge: Cambridge University Press.

———. 1992b. What Makes Tony Run? Schemas as Motives Reconsidered. In *Human Motives and Cultural Models*, ed. R. G. D'Andrade and C. Strauss, pp. 197–224. Cambridge: Cambridge University Press.

———. 1997. Partly Fragmented, Partly Integrated: An Anthropological Examination of "Postmodern Fragmented Subjects." *Cultural Anthropology* 12(3): 362–404.

Strauss, Claudia, and Naomi Quinn. 1994. A Cognitive/Cultural Anthropology. In *Assessing Cultural Anthropology*, ed. R. Borofsky, pp. 284–97. New York: McGraw-Hill.

———. 1997. *A Cognitive Theory of Cultural Meaning*. Cambridge: Cambridge University Press.

Stromberg, Peter. 1985. The Impression Point: Synthesis of Symbol and Self. *Ethos* 13(1): 56–74.

———. 1990. Ideological Language in the Transformation of Identity. *American Anthropologist* 92(1): 42–56.

———. 1993. *Language and Self-Transformation. A Study of the Christian Conversion Narrative*. Cambridge: Cambridge University Press.

Sultán, Madeleine. 1999. Choosing Islam: A Study of Swedish Converts. *Social Compass* 46(3): 325–35.

Svanberg, Ingvar, and David Westerlund, eds. 1999. *Blågul islam? Muslimer i Sverige.* Nora: Nya Doxa.
Svensson, Jonas. 1996. *Muslimsk feminism: Några exempel.* Religio 49. Skrifter utgivna av Teologiska Institutionen i Lund.
Wallace, Anthony F. C. 1961. *Culture and Personality.* New York: Random House.
Walzer, Michael. 1997. *On Toleration.* New Haven: Yale University Press.
Wikan, Unni. 1989. Managing the Heart to Brighten Face and Soul: Emotions in Balinese Morality and Health Care. *American Ethnologist* 16(2): 294–312.
———. 1995. The Self in a World of Urgency and Necessity. *Ethos* 23(3): 259–85.

Newspaper Articles

Dagens Nyheter, April 20, 1997.
Dagens Nyheter, January 20, 2002.
The Economist, August 28, 1999.
Sydsvenska Dagbladet, August 2, 1998.
Sydsvenska Dagbladet, August 21, 1998.
Sydsvenska Dagbladet, August 13, 1999.

Index